MACMILLAN/McGRAW-HILL
Math

Daily Practice Workbook with Summer Skills Refresher

Grade 4

The McGraw·Hill Companies

Macmillan
McGraw-Hill

Published by Macmillan/McGraw-Hill, of McGraw-Hill Education, a division of The McGraw-Hill Companies, Inc., Two Penn Plaza, New York, New York 10121.

Copyright © by Macmillan/McGraw-Hill. All rights reserved. No part of this publication may be reproduced or distributed in any form or by any means, or stored in a database or retrieval system, without the prior written consent of The McGraw-Hill Companies, Inc., including, but not limited to, network storage or transmission, or broadcast for distance learning.

Printed in the United States of America

6 7 8 9 066 08 07

Contents

Daily Practice

Lesson 1-1 Benchmark Numbers and Estimation 1
Lesson 1-2 Place Value Through Hundred Thousands 2
Lesson 1-3 Explore How Big Is a Million? 3
Lesson 1-4 Place Value Through Millions 4
Lesson 1-5 Compare and Order Numbers 5
Lesson 1-6 Problem Solving: Skill Using the Four-Step Process 6

Lesson 2-1 Count Money and Make Change 7
Lesson 2-2 Algebra: Compare and Order Money Amounts 8
Lesson 2-3 Round Numbers and Money 9
Lesson 2-4 Problem Solving: Strategy Make a Table 10

Lesson 3-1 Algebra: Addition and Subtraction Expressions 11
Lesson 3-2 Algebra: Properties of Addition 12
Lesson 3-3 Algebra: Addition Patterns 13
Lesson 3-4 Add Whole Numbers and Money 14
Lesson 3-5 Use Mental Math to Add 15
Lesson 3-6 Estimate Sums 16
Lesson 3-7 Problem Solving: Skill Estimate or Exact Answer 17

Lesson 4-1 Algebra: Explore Addition/Subtraction Equations 18
Lesson 4-2 Algebra: Subtraction Patterns 19
Lesson 4-3 Subtract Whole Numbers and Money 20
Lesson 4-4 Problem Solving: Strategy Write an Equation 21
Lesson 4-5 Use Mental Math to Subtract 22
Lesson 4-6 Estimate Differences 23
Lesson 4-7 Choose a Computation Method 24

Lesson 5-1 Tell Time 25
Lesson 5-2 Elapsed Time 26
Lesson 5-3 Calendar 27
Lesson 5-4 Range, Median, and Mode 28
Lesson 5-5 Collect and Organize Data 29
Lesson 5-6 Problem Solving: Skill Identify Extra and Missing Information 30

Lesson 6-1 Pictographs 31
Lesson 6-2 Bar Graphs 32
Lesson 6-3 Problem Solving: Strategy Use Logical Reasoning 33
Lesson 6-4 Coordinate Graphing 34
Lesson 6-5 Explore Making Line Graphs 35
Lesson 6-6 Interpreting Line Graphs 36
Lesson 6-7 Choose the Best Graph 37

Lesson 7-1 Algebra: Explore the Meaning of Multiplication 38
Lesson 7-2 Algebra: Properties of Multiplication 39
Lesson 7-3 Multiply by 2, 3, 4, and 6 40
Lesson 7-4 Multiply by 5, 7, 8, 9, and 10 41
Lesson 7-5 Problem Solving: Skill Choose an Operation 42
Lesson 7-6 Explore Square Numbers 43
Lesson 7-7 Algebra: Multiplication Table and Patterns 44

Lesson 8-1 Algebra: Explore the Meaning of Division 45
Lesson 8-2 Algebra: Relate Multiplication and Division 46
Lesson 8-3 Divide by 2 Through 12 47
Lesson 8-4 Algebra: Missing Factors 48
Lesson 8-5 Problem Solving: Strategy Work Backward 49
Lesson 8-6 Algebra: Expressions and Equations 50

Lesson 9-1 Algebra: Patterns and Properties 51
Lesson 9-2 Explore Multiplying by 1-Digit Numbers 52

Lesson 9-3 Multiply by 1-Digit Numbers...53
Lesson 9-4 Estimating Products..54
Lesson 9-5 Problem Solving: Skill Use an Overestimate or Underestimate 55

Lesson 10-1 Multiplying Greater Numbers...56
Lesson 10-2 Multiply Using Mental Math..57
Lesson 10-3 Choose a Computation Method...58
Lesson 10-4 Algebra: Functions..59
Lesson 10-5 Algebra: Graphing Functions...60
Lesson 10-6 Problem Solving: Strategy Find a Pattern..............................61

Lesson 11-1 Algebra: Patterns of Multiplication...................................62
Lesson 11-2 Multiply by Multiples of 10...63
Lesson 11-3 Explore Multiplying by 2-Digit Numbers................................64
Lesson 11-4 Multiply by 2-Digit Numbers...65
Lesson 11-5 Problem Solving: Skill Solve Multistep Problems.......................66

Lesson 12-1 Estimate Products...67
Lesson 12-2 Multiplying Greater Numbers...68
Lesson 12-3 Problem Solving: Strategy Make a Graph................................69
Lesson 12-4 Multiply Using Mental Math..70
Lesson 12-5 Choose a Computation Method...71

Lesson 13-1 Algebra: Division Patterns..72
Lesson 13-2 Estimate Quotients..73
Lesson 13-3 Explore Dividing by 1-Digit Numbers...................................74
Lesson 13-4 Divide by 1-Digit Numbers...75
Lesson 13-5 Problem Solving: Skill Interpret the Remainder........................76
Lesson 13-6 Quotients with Zeros..77

Lesson 14-1 Divide Greater Numbers..78
Lesson 14-2 Choose a Computation Method...79
Lesson 14-3 Problem Solving: Strategy Guess and Check.............................80
Lesson 14-4 Find the Better Buy...81
Lesson 14-5 Explore Finding the Mean..82
Lesson 14-6 Find the Mean...83

Lesson 15-1 Algebra: Division Patterns..84
Lesson 15-2 Estimating Quotients..85
Lesson 15-3 Divide 2-Digit Numbers by Multiples of 10.............................86
Lesson 15-4 Explore Dividing by 2-Digit Numbers...................................87
Lesson 15-5 Divide by 2-Digit Numbers...88
Lesson 15-6 Problem Solving: Skill Use an Overestimate or Underestimate...........89

Lesson 16-1 Adjusting Quotients...90
Lesson 16-2 Choose a Computation Method...91
Lesson 16-3 Problem Solving: Strategy Choose a Strategy...........................92
Lesson 16-4 Algebra: Order of Operations..93

Lesson 17-1 Explore Nonstandard Length, Width, and Height.........................94
Lesson 17-2 Explore Customary Length to $\frac{1}{4}$ Inch........................95
Lesson 17-3 Customary Capacity and Weight...96
Lesson 17-4 Algebra: Convert Customary Units......................................97
Lesson 17-5 Problem Solving: Skill Check for Reasonableness.......................98

Lesson 18-1 Explore Metric Length...99
Lesson 18-2 Metric Capacity and Mass...100
Lesson 18-3 Algebra: Convert Metric Units..101
Lesson 18-4 Problem Solving: Strategy Logical Reasoning..........................102
Lesson 18-5 Temperature..103

Lesson 19-1 3-Dimensional Figures..104
Lesson 19-2 2-Dimensional Figures..105
Lesson 19-3 Lines, Line Segments, and Rays.......................................106
Lesson 19-4 Angles...107

Lesson 19-5 Triangles and Quadrilaterals ... 108
Lesson 19-6 Problem Solving: Skill Use a Diagram ... 109
Lesson 19-7 Parts of a Circle ... 110

Lesson 20-1 Congruent and Similar ... 111
Lesson 20-2 Transformations and Symmetry ... 112
Lesson 20-3 Problem Solving: Strategy Find a Pattern ... 113
Lesson 20-4 Perimeter ... 114
Lesson 20-5 Circumference ... 115
Lesson 20-6 Algebra: Area ... 116
Lesson 20-7 Algebra: Explore Volume ... 117

Lesson 21-1 Parts of a Whole ... 118
Lesson 21-2 Parts of a Group ... 119
Lesson 21-3 Find Equivalent Fractions and Fractions in Simplest Form ... 120
Lesson 21-4 Algebra: Compare and Order Fractions ... 121
Lesson 21-5 Problem Solving: Skill Check for Reasonableness ... 122
Lesson 21-6 Explore Finding Parts of a Group ... 123
Lesson 21-7 Mixed Numbers ... 124

Lesson 22-1 Likely and Unlikely ... 125
Lesson 22-2 Explore Probability ... 126
Lesson 22-3 Problem Solving: Strategy Draw a Tree Diagram ... 127
Lesson 22-4 Explore Making Predictions ... 128

Lesson 23-1 Add Fractions with Like Denominators ... 129
Lesson 23-2 Explore Adding Fractions with Unlike Denominators ... 130
Lesson 23-3 Add Fractions with Unlike Denominators ... 131
Lesson 23-4 Problem Solving: Skill Choose an Operation ... 132

Lesson 24-1 Subtract Fractions with Like Denominators ... 133
Lesson 24-2 Explore Subtracting Fractions with Unlike Denominators ... 134
Lesson 24-3 Subtract Fractions with Unlike Denominators ... 135
Lesson 24-4 Problem Solving: Strategy Solve a Simpler Problem ... 136
Lesson 24-5 Circle Graphs ... 137

Lesson 25-1 Explore Fractions and Decimals ... 138
Lesson 25-2 Tenths and Hundredths ... 139
Lesson 25-3 Problem Solving: Skill Choose a Representation ... 140
Lesson 25-4 Thousandths ... 141

Lesson 26-1 Decimals Greater than 1 ... 142
Lesson 26-2 Algebra: Compare and Order Decimals ... 143
Lesson 26-3 Problem Solving: Strategy Draw a Diagram ... 144
Lesson 26-4 Round Decimals ... 145

Lesson 27-1 Explore Adding Decimals ... 146
Lesson 27-2 Add Decimals ... 147
Lesson 27-3 Problem Solving: Skill Choose an Operation ... 148
Lesson 27-4 Estimate Sums ... 149
Lesson 27-5 Choose a Computation Method ... 150

Lesson 28-1 Explore Subtracting Decimals ... 151
Lesson 28-2 Subtract Decimals ... 152
Lesson 28-3 Problem Solving: Strategy Act It Out ... 153
Lesson 28-4 Estimate Differences ... 154
Lesson 28-5 Choose a Computation Method ... 155

Summer Skills Refresher
Number Sense, Concepts, and Operations ... 159
Measurement ... 161
Geometry and Spatial Sense ... 163
Algebraic Thinking ... 165
Data Analysis and Probability ... 167

Name _____

Explore: How Big Is a Million?

Find the answer.

1. How many 10-by-10 grids would you need to make a thousand cube?

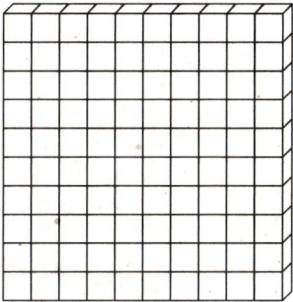

_____

2. How many thousand cubes would you need to make a million?

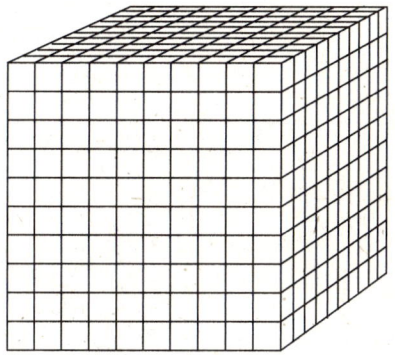

_____

3. How many hundreds are in 1,000? _____

4. How many hundreds are in 10,000? _____

5. How many thousands are in 1,000? _____

6. How many thousands are in 10,000? _____

7. How many thousands are in 100,000? _____

8. How many thousands are in 1,000,000? _____

9. How many ten thousands are in 10,000? _____

10. How many ten thousands are in 100,000? _____

11. How many ten thousands are in 1,000,000? _____

12. How many hundred thousands are in 100,000? _____

13. How many hundred thousands are in 1,000,000? _____

Use with Grade 4, Chapter 1, Lesson 3, pages 6–7.

Name _____

Place Value Through Millions

Write the expanded form and the word form for each number.

1. 1,420,316 _____

2. 2,672,400 _____

3. 12,060,072 _____

4. 785,004,012 _____

Write the value of each underlined digit.

5. <u>8</u>42,753 _____   6. <u>6</u>,782,141 _____

7. 1<u>5</u>3,428,090 _____   8. <u>7</u>15,124,068 _____

Write each number in standard form.

9. one million, two hundred thousand, five _____

10. thirty-eight million, four hundred thousand, eight _____

11. five hundred eighty million, sixty-two thousand, seventeen _____

12. two hundred fifty-four million, seven thousand, five _____

Algebra Find the missing number.

13. 42,865 = 40,000 + _____ + 800 + 60 + 5

14. 168,943 = 100,000 + 60,000 + 8,000 + _____ + 40 + 3

15. 888,888 = 800,000 + _____ + 8,000 + 800 + 80 + 8

4   Use with Grade 4, Chapter 1, Lesson 4, pages 8–9.

Name: Ashtin

Compare and Order Numbers +25

1-5 PRACTICE

Compare. Write >, <, or =.

1. 3,874 > 3,862
2. 5,741 < 5,862
3. 7,824 > 7,724
4. 14,624 > 1,462
5. 42,542 > 41,617
6. 32,145 < 32,264
7. 10,142 < 12,641
8. 25,632 = 25,632
9. 89,000 > 87,999
10. 150,420 > 100,042
11. 434,121 > 432,154
12. 187,654 < 197,541
13. 782,421 > 782,342
14. 642,134 = 642,134
15. 874,158 < 972,421

Order from greatest to least.

16. 3,421; 3,641; 3,481; 3,562   3,641; 3,562; 3,481; 3,421
17. 21,649; 21,842; 20,649   21,842; 21,649; 20,649
18. 72,642; 71,848; 70,621   72,642; 71,848; 70,621
19. 748,629; 747,832; 748,532   748,629; 748,582; 747,832

Order from least to greatest.

20. 6,421; 6,878; 8,768; 6,543   6,421; 6,543; 6,878; 8,768
21. 25,421; 24,462; 24,416   24,416; 24,462; 25,421
22. 324,621; 324,742; 325,697   324,621; 324,742; 325,697
23. 524,607; 525,712; 524,872   524,607; 524,872; 525,712

Problem Solving

Solve.

24. Sean has 1,575 bird stamps and Li has 2,075 bird stamps. Cindy has a number of stamps between Sean's and Li's numbers. Is the number 1,075 or 1,755? Explain.

1,755, 1,075 is less than 1,575

25. Highway A is 1,275 miles long and Highway B is 1,850 miles long. Highway C is the longest of the three. Is the number 1,875 miles or 1,175 miles? Explain.

1,875 is longer than 1,850

Use with Grade 4, Chapter 1, Lesson 5, pages 10–12.

5

Name _____

Problem Solving: Skill
Using the Four-Step Process

Solve. Use the four-step process.

1. A marlin can move at a speed of 50 miles per hour. A striped dolphin can move 19 miles per hour. A killer whale can move 55 miles per hour. List the animals in order from slowest to fastest.

   _____

   _____

2. Brandon, Timothy, and Norah have pet care services. Last year, Brandon earned $712, Timothy earned $1,110, and Norah earned $650. List the business owners in order from who earned the greatest amount to who earned the least amount.

   _____

3. A poll shows that 311 students have dogs, 424 students have cats, 96 students have birds, and 38 students have a different pet. Which kind of pet is owned by the most students?

   _____

4. The pet shelter takes in 24 dogs in April, 41 dogs in May, and 39 dogs in June. List the months in order, beginning with the month in which the shelther took in the fewest dogs.

   _____

5. Dylan spots 48 birds. Nicole spots 51 birds. Who spots fewer birds?

   _____

6. In 2002, about 36,000,000 people visited aquariums and about 86,000,000 people visited zoos. Did more people visit aquariums or zoos?

   _____

7. A bottle-nosed dolphin can weigh up to 440 pounds. A common dolphin can weigh up to 165 pounds. Which kind of dolphin is likely to be heavier?

   _____

8. On Friday, 660 people go to Ocean National Park. On Saturday, 1,096 people go to the park. On Sunday, 998 people go. On which day did the most people go to the park?

   _____

Use with Grade 4, Chapter 1, Lesson 6, pages 14–15.

Name **Ashtin**

Count Money and Make Change +16

2-1 PRACTICE

Write the amount of money shown.

1. 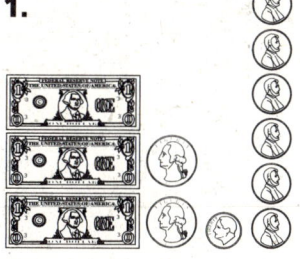 2.  3.

$3.66 $7.82 $9.70

Tell which coins and bills make the amount shown below.

4. $0.89 _eight dimes, one nickel, and four pennies_

5. $3.62 _three one dollar bills, six dimes, and two pennies_

6. $7.67 _one five dollar bill, two one dollar bills, six dimes, one nickel and two pennies._

Find the amount of change.

7. Cost: $0.59
   Amount given: $1.00
   41¢

8. Cost: $2.45
   Amount given: $5.00
   $2.55

9. Cost: $7.81
   Amount given: $10.00
   $2.19

10. Cost: $0.86
    Amount given: $5.00
    $4.14

11. Cost: $3.09
    Amount given: $10.00
    $6.91

12. Cost: $9.25
    Amount given: $10.00
    10.75¢

Algebra Find each missing number.

13. $7.50; $7.60; _$7.70_; $7.80

14. $25.95; _$30.95_; $35.95; $40.95

Problem Solving
Solve.

15. Andy gives the cashier $5.00 to pay for a $3.75 calendar. How much change does he receive?
    $1.25

16. Lowanda receives 1 quarter, 2 dimes, and 1 nickel in change. How much money did she get?
    50¢

Use with Grade 4, Chapter 2, Lesson 1, pages 20–22.

7

Name _____

Compare and Order Money Amounts • Algebra 2-2 PRACTICE

Compare. Write >, <, or =.

1. $0.33 ◯ $0.03
2. $0.73 ◯ $0.37
3. $0.92 ◯ $9.20
4. $45.16 ◯ $45.12
5. $0.09 ◯ $9.00
6. $4.73 ◯ $4.70
7. $67.95 ◯ $66.89
8. $30.72 ◯ $31.04
9. $55.91 ◯ $55.19
10. $127.43 ◯ $126.50
11. $275.33 ◯ $280.21
12. $360.44 ◯ $359.99
13. $710.03 ◯ $711.87
14. $549.36 ◯ $621.33
15. $852.93 ◯ $852.91

Order from greatest to least.

16. $0.55; $1.59; $0.56 _____
17. $2.75; $0.98; $1.00 _____
18. $43.89; $43.98; $43.79 _____
19. $104.62; $95.50; $111.24 _____

Order from least to greatest.

20. $0.59; $0.09; $0.90 _____
21. $45.88; $51.72; $33.66 _____
22. $106.45; $93.88; $102.29 _____
23. $688.02; $684.97; $688.53 _____

Problem Solving
Solve.

24. Carlos pays $3.75 for a box of 48 crayons. Ellie pays $3.95 for the same box at a different store. Who pays more for the box of crayons? Explain.

_____

25. Al's stamp album costs $12.75 and Li's album costs $18.50. Cindy's album costs the most. Did it cost $18.75 or $11.75? Explain.

_____

8   Use with Grade 4, Chapter 2, Lesson 2, pages 24–25.

Name _____

Addition and Subtraction Expressions • Algebra

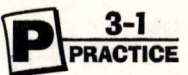

Find the value of the expression.

1. $9 - y$ for $y = 2$ _____
2. $m + 3$ for $m = 2$ _____
3. $3 + x$ when $x = 10$ _____
4. $12 - w$ when $w = 4$ _____
5. $z + 37$ when $z = 29$ _____
6. $54 + p$ when $p = 3$ _____
7. $71 - l$ when $l = 29$ _____
8. $k + 33$ when $k = 48$ _____
9. $p - 109$ when $p = 275$ _____
10. $288 + n$ when $n = 106$ _____
11. $121 + g$ when $g = 129$ _____
12. $500 - t$ when $t = 266$ _____

Write an expression for each situation.

13. 7 more than x _____
14. 5 and p more _____
15. 2 and m more _____
16. 3 more than g _____
17. 12 and y more _____
18. 25 and b more _____
19. 155 more than q _____
20. 341 and f more _____

Write an expression for the pattern.

21. $10 + 1, 10 + 2, 10 + 3, \ldots$ _____
22. $45 - 5, 45 - 6, 45 - 7, \ldots$ _____
23. $62 + 3, 62 + 4, 62 + 5, \ldots$ _____

Problem Solving
Solve.

24. George earns $30 plus tips each day. Write an expression to show his total daily pay. If George received $8 in tips yesterday, how much did he earn in all?

25. Tanesha has 24 marbles. She gives away x number of marbles. Write an expression for the number of marbles she has left.

_____ _____

Use with Grade 4, Chapter 3, Lesson 1, pages 46–47.

11

Name_____

Properties of Addition • Algebra

Find the value of the variable.

1. $a + 5 = 10$ _____
2. $12 + x = 20$ _____
3. $n + 4 = 8$ _____
4. $l + 0 = 6$ _____
5. $t + 12 = 15$ _____
6. $11 + b = 17$ _____

Find the sum or difference. Write the related number sentences.

7. $2 + 9 = $ _____
8. $35 + 4 = $ _____
9. $54 + 0 = $ _____

_____ _____ _____
_____ _____ _____
_____ _____ _____

Write the related number sentences for each set of numbers.

10. 4, 5, 9
11. 11, 24, 35
12. 0, 46, 46

_____ _____ _____
_____ _____ _____
_____ _____ _____

13. 17, 18, 35
14. 6, 42, 48
15. 30, 50, 80

_____ _____ _____
_____ _____ _____
_____ _____ _____

Problem Solving
Solve.

16. Ken has 6 coins in his collection. Barb has 5 more coins than Ken. How many coins does Barb have?

_____

17. Meg has 13 coins in her collection. Then she gives 7 coins to her cousin. How many coins does Meg have now?

_____

Use with Grade 4, Chapter 3, Lesson 2, pages 48–50.

Name_____

Addition Patterns • Algebra

Write the number that makes each sentence true.

1. $8 + 8 = n$ _____
    $80 + 80 = n$ _____
    $800 + 800 = n$ _____
    $8,000 + 8,000 = n$ _____
    $80,000 + 80,000 = n$ _____
    $800,000 + 800,000 = n$ _____

2. $7 + 6 = n$ _____
    $70 + 60 = n$ _____
    $700 + 600 = n$ _____
    $7,000 + 6,000 = n$ _____
    $70,000 + 60,000 = n$ _____
    $700,000 + 600,000 = n$ _____

3. $5 + 9 = n$ _____
    $50 + 90 = n$ _____
    $500 + 900 = n$ _____
    $5,000 + 9,000 = n$ _____
    $50,000 + 90,000 = n$ _____
    $500,000 + 900,000 = n$ _____

4. $8 + 9 = n$ _____
    $80 + 90 = n$ _____
    $800 + 900 = n$ _____
    $8,000 + 9,000 = n$ _____
    $80,000 + 90,000 = n$ _____
    $800,000 + 900,000 = n$ _____

Add mentally.

5. $500 + 400 =$ _____
6. $3,000 + $ _____ $=$ _____
7. $30,000 + 80,000 =$ _____
8. $700 + 800 =$ _____
9. $600 + 500 =$ _____
10. $70,000 + 30,000 =$ _____
11. $100,000 + 900,000 =$ _____
12. $800,000 + 500,000 =$ _____

Problem Solving
Solve.

13. A music store made $50,000 selling CDs and tapes in December. In January, the store made $30,000. How much did the store make in all?

    _____

14. The Green Hornets sold 800,___ copies of their first CD. They s___ 500,000 copies of their second ___ How many CDs did the Green Hor___ sell in all?

    _____

Use with Grade 4, Chapter 3, Lesson 3, pages 52–53.

Name_____

Add Whole Numbers and Money

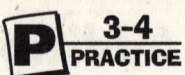

Find each sum.

1. 688
   + 207

2. 574
   + 434

3. 757
   + 529

4. $8.72
   + 1.38

5. $2.98
   + 0.59

6. 989
   + 624

7. 8,489
   + 2,467

8. $3,824
   +   962

9. 5,174
   +   327

10. $12.57
   +   7.43

11. 6,672
   + 878

12. $78.29
   + 45.32

13. 12,345
   + 67,890

14. 43,802
   + 7,526

15. 24,316
   +   893

16. 183,462
   + 570,184

17. 3,421
     1,657
   +   728

18. 24,177
       410
   + 4,586

19. 341,249
     85,278
   + 203,655

20. $275.35
      62.80
   +   9.82

21. $7.77 + $6.66 = _____

22. 7,709 + 3,047 = _____

Algebra Find each sum. Use properties to help you.

23. 432 + 215 + 308 = _____

24. 5,780 + 750 + 130 = _____

Problem Solving
Solve.

25. At the Lakeside School, 522 students ride the bus and 714 students walk or are driven to school. How many students attend Lakeside School?

26. Last week, $325 worth of play tickets and $729 worth of carnival tickets were sold. How much money was collected altogether?

_____

14

Use with Grade 4, Chapter 3, Lesson 4, pages 54–56.

Name _____

Use Mental Math to Add

Add mentally.

1. 32 + 45 = _____
2. 21 + 64 = _____
3. 35 + 13 = _____
4. $39 + $24 = _____
5. 48 + 31 = _____
6. 298 + 311 = _____
7. 595 + 409 = _____
8. 255 + 344 = _____
9. 238 + 495 = _____
10. 730 + 214 = _____
11. 891 + 108 = _____
12. $256 + $222 = _____
13. 4,524 + 3,173 = _____
14. 8,999 + 1,333 = _____
15. 2,295 + 2,124 = _____
16. 1,487 + 1,511 = _____

Algebra Write the value of each missing number.

17. $36 + a = 86$ _____
18. $b + 61 = 81$ _____
19. $\$498 + c = \$698$ _____
20. $d + 298 = 598$ _____
21. $e + 657 = 957$ _____
22. $\$63 + h = \$243$ _____
23. $\$725 + k = \$1{,}125$ _____
24. $m + 837 = 1{,}137$ _____
25. $1{,}650 + n = 3{,}300$ _____
26. $r + \$750 = \$1{,}500$ _____

Problem Solving
Solve.

27. There are 38 dogs and 24 cats at the pet show. How many cats and dogs are there in all?

_____

28. The pet show committee spends $316 on dog treats and $299 on cat treats. How much does the committee spend on treats?

_____

Use with Grade 4, Chapter 3, Lesson 5, pages 58–59.

15

Estimate Sums

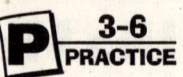

Estimate each sum. Show your work.

1. 478 + 597 _____
2. $8.65 + $7.15 _____
3. $0.32 + $0.65 _____
4. 4,990 + 405 _____
5. 2,188 + 5,621 _____
6. 47,522 + 3,721 _____
7. 863,122 + 254,087 _____

Add. Estimate to check if your answer is reasonable.

8. 621 + 308 = _____   9. 2,188 + 5,621 = _____

10. $4.20 + $8.12 = _____   11. 601,128 + 328,125 = _____

Compare. Write > or < to make a true sentence.

12. 176 + 335 ◯ 400   13. 243 + 50 ◯ 300
14. 500 ◯ 251 + 127   15. 900 ◯ 895 + 68
16. 1,348 + 2,489 ◯ 5,000   17. 4,725 + 321 ◯ 3,923 + 289
18. 9,000 ◯ 4,487 + 5,672   19. 8,000 ◯ 6,081 + 950
20. 22,152 + 28,174 ◯ 60,000   21. 49,912 + 2,839 ◯ 5,000

Problem Solving
Solve.

22. Julio wants to buy drawing paper for $8.50 and brushes for $19.95. About how much will he spend?

23. The fourth-grade students make 268 posters about bicycle safety. The fifth-grade students make 229. About how many posters do the students make altogether?

_____   _____

16   Use with Grade 4, Chapter 3, Lesson 6, pages 60–61.

Name _____

Problem Solving: Skill
Estimate or Exact Answer

Solve. Explain why you gave an estimate or exact answer.

1. James, Max, and Melba collect baseball cards. James has 870 cards, Max has 569 cards, and Melba has 812 cards. Do the three friends have more than 2,000 baseball cards?

   _____

   _____

2. Nicki has a collection of 79 shells and 64 rocks. How many items are in her collection?

   _____

   _____

3. Kelly has a coin collection. Her quarters are worth $104.50. Her dimes are worth $75.10. Her nickels are worth $27.75. What is the total value of Kelly's coin collection?

   _____

   _____

4. The Comic Book Show sells 474 tickets on Friday and 396 tickets on Saturday. About how many tickets does the Comic Book Show sell?

   _____

   _____

Mixed Strategy Review
Use data from the table for problems 5–6.

5. How many people visited the museum on Saturday and Sunday?

   _____

6. About how many people visited the museum on Wednesday, Thursday, and Friday?

   _____

Museum Visitors	
Wednesday	377
Thursday	405
Friday	529
Saturday	836
Sunday	915

Use with Grade 4, Chapter 3, Lesson 7, pages 62–63.

17

Name _____

Explore Addition and Subtraction Equations • Algebra

P PRACTICE 4-1

Use models to find the value of the variable.

1. $x + 7 = 9$ $x = $ _____

2. $p - 8 = 4$ $p = $ _____

3. $s - 2 = 1$ $x = $ _____

4. $b + 6 = 11$ $p = $ _____

Solve each equation. Check your answer.

5. $2 + v = 4$ $v = $ _____

6. $d + 8 = 13$ $d = $ _____

7. $s - 5 = 2$ $s = $ _____

8. $11 + x = 15$ $x = $ _____

9. $7 + p = 10$ $p = $ _____

10. $b - 7 = 11$ $b = $ _____

11. $l + 3 = 5$ $l = $ _____

12. $g - 5 = 6$ $g = $ _____

13. $a - 1 = 12$ $a = $ _____

14. $u + 0 = 10$ $u = $ _____

15. $m - 8 = 1$ $m = $ _____

16. $h - 2 = 17$ $h = $ _____

17. $c + 4 = 16$ $c = $ _____

18. $12 + k = 13$ $k = $ _____

19. $f + 9 = 19$ $f = $ _____

20. $n - 0 = 0$ $n = $ _____

21. $r - 4 = 10$ $r = $ _____

22. $z + 16 = 16$ $z = $ _____

18   Use with Grade 4, Chapter 4, Lesson 1, pages 68–69.

Name _____

Subtraction Patterns • Algebra

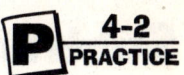

Write the number that makes each sentence true.

1. 12 − 8 = n _____
   120 − 80 = n _____
   1,200 − 800 = n _____
   12,000 − 8,000 = n _____
   120,000 − 80,000 = n _____
   1,200,000 − 800,000 = n _____

2. 16 − 7 = n _____
   160 − 70 = n _____
   1,600 − 700 = n _____
   16,000 − 7,000 = n _____
   160,000 − 70,000 = n _____
   1,600,000 − 700,000 = n _____

3. 11 − 5 = n _____
   110 − 50 = n _____
   1,100 − 500 = n _____
   11,000 − 5,000 = n _____
   110,000 − 50,000 = n _____
   1,100,000 − 500,000 = n _____

4. 15 − 8 = n _____
   150 − 80 = n _____
   1,500 − 800 = n _____
   15,000 − 8,000 = n _____
   150,000 − 80,000 = n _____
   1,500,000 − 800,000 = n _____

Subtract mentally.

5. 1,200 − 600 = _____
6. $8,000 − $3,000 = _____
7. 600,000 − 500,000 = _____
8. 70,000 − 50,000 = _____
9. $13,000 − $9,000 = _____
10. 160,000 − 80,000 = _____
11. 140,000 − 50,000 = _____
12. 1,200,000 − 600,000 = _____

Problem Solving
Solve.

13. A video store rented 900,000 videos last year. This year the store rented 1,500,000 videos. How many more videos did it rent this year?

    _____

14. The price for a house is $120,000. Ms. Smith decides to make an offer that is $30,000 less than the price. How much does Ms. Smith offer for the house?

    _____

Use with Grade 4, Chapter 4, Lesson 2, pages 70–71.

Name_____

Subtract Whole Numbers and Money

4-3 PRACTICE

Subtract. Check by adding.

1. 757
   − 28

2. $582
   − 492

3. 693
   − 516

4. 600
   − 58

5. $2.48
   − 1.95

6. 2,345
   − 1,658

7. $67.89
   − 18.95

8. $11,321
   − 979

9. 8,000
   − 2,450

10. 3,523
    − 2,846

11. $33,572
    − 13,689

12. 74,125
    − 65,239

13. 49,785
    − 8,998

14. 98,142
    − 617

15. $224.20

16. $4,561.71
    − 291.68

17. 389,243
    − 136,354

18. $900,000
    − 98,276

19. 914,617
    − 117,814

20. $7,211.53
    − 5,926.84

21. 500 − 124 = _____

22. $9.12 − $7.58 = _____

23. 42,625 − 9,846 = _____

24. 70,000 − 52,087 = _____

25. $311.42 − $4.65 = _____

26. $578,423 − $89,743 = _____

27. (276,410 + 39,257) − 6,413 = _____

Problem Solving
Solve.

28. A toy factory made 32,154 board games on Monday. On Tuesday it made 31,687 board games. How many more board games did the factory make on Monday?

29. A store earned $12,415 selling puzzles this week. Last week it earned $9,326 selling puzzles. How much more did the store earn this week?

20   Use with Grade 4, Chapter 4, Lesson 3, pages 72–74.

Name _____

Problem Solving: Strategy
Write an Equation

Write an equation to solve.

1. Meg buys candle-making supplies for $37. She has $25 left. How much money did Meg have before she bought the supplies?

   _____

2. Sally has finished 86 squares in her quilt. The quilt will have 100 squares. How many squares does Sally still have to make?

   _____

3. Eric sells a painting for $125. He sells a sculpture for $390. How much money does Eric earn in all?

   _____

4. Noah has saved $42. How much more money does he need to buy a rare coin for $90?

   _____

Mixed Strategy Review
Solve. Use any strategy.

5. Howard has 75 shells. On a trip, he collects another 16 shells. How many shells does he have now?

   _____

   Strategy: _____

6. Tom makes letters for a sign that says "Arts and Crafts Fair." Which letter does Mark need to make the most of?

   _____

   Strategy: _____

7. **Social Studies** During the 1800s, sailors made carvings called scrimshaw on whale teeth, whalebone, and tortoise shells. Suppose a sailor made a carving in 1805. A collector buys the carving in 2000. How many years old is the carving?

   _____

   Strategy: _____

8. **Write a problem** that you could write an equation to solve. Share it with others.

   _____

   _____

   _____

Use with Grade 4, Chapter 4, Lesson 4, pages 76–77.

21

Name_____

Use Mental Math to Subtract

4-5 PRACTICE

Subtract mentally.

1. 46 − 7 = _____
2. 81 − 36 = _____
3. 53 − 19 = _____
4. 99 − 19 = _____
5. $78 − $49 = _____
6. 92 − 28 = _____
7. 74 − 38 = _____
8. 95 − 37 = _____
9. 64 − 37 = _____
10. 687 − 48 = _____
11. $273 − $58 = _____
12. 394 − 86 = _____
13. $704 − $589 = _____
14. 745 − 597 = _____
15. 782 − 203 = _____
16. 613 − 309 = _____
17. 555 − 299 = _____
18. 998 − 145 = _____
19. 578 − 465 = _____

Write the number that makes each equation true.

20. $648 - a = 548$ _____
21. $b - 60 = 340$ _____
22. $c - 412 = 388$ _____
23. $d - 235 = 665$ _____
24. $950 - e = 400$ _____
25. $823 - h = 123$ _____
26. $k - 599 = 301$ _____
27. $450 - m = 100$ _____
28. $775 - n = 200$ _____
29. $r - 300 = 1{,}456$ _____

Problem Solving
Solve.

30. Josh buys a wooden horse for $4.89. He gives the cashier $5.00. How much change should Josh receive?

_____

31. A bicycle shop has 309 water bottles in stock. Ashley buys 259 bottles. How many bottles are left?

_____

Name_____

Estimate Differences

Estimate each difference. Show your work.

1. 467 − 215 _____

2. 2,835 − 1,487 _____

3. $13.95 − $7.25 _____

4. 65,074 − 15,472 _____

5. 174,921 − 18,421 _____

Subtract. Estimate to check that each answer is reasonable.

6.	7.	8.	9.	10.
835	$81.79	6,984	242,003	654,026
− 487	− 31.55	− 322	− 49,887	− 529,620

11. $0.88 − $0.35 = _____   12. 787,008 − 117,584 = _____

Algebra Compare. Write > or < to make the sentence true.

13. 4,173 − 2,589 ◯ 2,000   14. 8,329 − 957 ◯ 7,000

15. $300.00 ◯ $367.20 − $59.45   16. 600 ◯ 938 − 452

17. 15,425 − 3,535 ◯ 10,000   18. 8,053 − 7,645 ◯ 1,000

19. 42,345 − 16,174 ◯ 20,000   20. 48,592 − 961 ◯ 4,000

Problem Solving
Solve.

21. Last year, 787,897 copies of *Science Monthly* were sold. This year, 914,632 copies were sold. About how many more copies were sold this year than last year?

    _____

22. The Hoop Store spends $129.99 for an ad in *Science Monthly*. The store spends $19.29 for an ad in the *Allentown News*. About how much more does the store spend on advertising in *Science Monthly* than in the *Allentown News*?

    _____

Use with Grade 4, Chapter 4, Lesson 6, pages 80–81.

23

Name _____

Choose a Computation Method

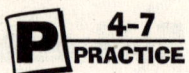

Add or subtract. Tell which method you used.

1. 375
   + 624
   _____

2. 894
   − 695
   _____

3. 2,900
   + 2,100
   _____

4. 3,799
   − 1,799
   _____

5. $74.66
   + 35.91
   _____

6. $274.88
   −   99.19
   _____

7. 7,991 − 6,382 = _____

8. 200 + 15 + 85 = _____

9. $150 + $230 = _____

10. 388,261 − 68,937 = _____

Problem Solving
Solve.

11. There were 25,899 people at the baseball game on Saturday. There were 1,997 more people at the game on Sunday than on Saturday. How many people were at the game on Sunday?

_____

12. There were 18,362 people at the basketball game on Wednesday. There were 563 fewer people at the game on Friday than on Wednesday. How many people were at the game on Friday?

_____

24

Use with Grade 4, Chapter 4, Lesson 7, pages 82–83.

Name _____

Tell Time

Write the time in two ways.

1.

2.

3.

_____   _____   _____

_____   _____   _____

Tell how much time.

4. 120 minutes = ☐ hours

5. ☐ seconds = 3 minutes

6. $\frac{1}{2}$ hour = ☐ minutes

7. 15 minutes = ☐ hour

8. ☐ minutes = $2\frac{1}{2}$ hours

9. ☐ minutes = $1\frac{1}{4}$ hours

Algebra Describe and complete the conversion patterns.

10.
Minutes	60	120	180	240	300
Hours	1	2	☐	☐	☐

_____

11.
Minutes	1	2	3	4	5
Seconds	60	120	☐	☐	☐

_____

12. Debbie spends 45 minutes at the dentist. What part of an hour does she spend at the dentist?

_____

13. Ben spends $5\frac{1}{2}$ hours in school. Does he spend more or less than 300 minutes in school? Explain.

_____

Use with Grade 4, Chapter 5, Lesson 1, pages 98–100.

Name_____

Elapsed Time

How much time has passed?

1. Begin: 12:00 P.M.
   End: 2:20 P.M.
   _____

2. Begin: 1:15 A.M.
   End: 1:50 A.M.
   _____

3. Begin: 11:05 P.M.
   End: 1:00 A.M.
   _____

4. Begin: 2:25 A.M.
   End: 5:40 A.M.
   _____

5. Begin: 3:40 P.M.
   End: 12:00 A.M.
   _____

6. Begin: 5:45 A.M.
   End: 12:15 P.M.
   _____

7. Begin: 8:10 P.M.
   End: 1:55 A.M.
   _____

8. Begin: 9:30 A.M.
   End: 2:10 P.M.
   _____

9. Begin: 10:35 P.M.
   End: 8:00 A.M.
   _____

What time will it be in 1 hour 20 minutes?

10.

11.

12.

_____   _____   _____

Write the missing numbers.

13. 5:16 A.M. is _____ minutes after 5:00 A.M.

14. 2:45 P.M. is _____ minutes before 3:00 P.M.

15. 7:22 P.M. is _____ hours _____ minutes after 7:00 P.M.

16. 9:58 A.M. is _____ minutes before _____ A.M.

Problem Solving
Solve.

17. Lisa leaves her house at 8:45 A.M. She gets to karate class 35 minutes later. At what time does Lisa get to karate class?

18. The Big Beach bus leaves the city at 6:40 P.M. The bus arrives at the beach at 8:25 P.M. How long is the trip to the beach?

Name _____

Calendar

Use the calendars for July and August for problems 1–8.

July 2000

S	M	T	W	T	F	S
						1
2	3	4 Independence Day	5	6	7	8
9	10	11	12	13	14	15
16	17	18	19	20	21	22
23	24	25	26	27	28	29
30	31					

August 2000

S	M	T	W	T	F	S
		1	2 Nick arrives	3	4	5
6	7	8	9	10	11	12
13	14	15 Football practice begins	16	17	18	19
20	21	22	23	24	25	26
27	28	29	30	31		

1. What is the date of the fourth Thursday in July?

   _____

2. On what day of the week is Independence Day?

   _____

3. Cindy will return from vacation on the Monday after Nick arrives. On what date will Cindy return?

   _____

4. If soccer camp runs from July 7 through the following Saturday, how long is soccer camp?

   _____

5. Justin is moving to a new town on August 1. The movers are coming 4 days before that. On what date will the movers arrive?

   _____

6. Jason has a violin lesson every Wednesday. How many lessons will he have in July and August?

   _____

7. Nick will leave on August 30. For how many weeks will he visit?

   _____

8. Pat saw the dentist on July 25. He has another appointment 10 days later. On what date is Pat's appointment?

   _____

Use with Grade 4, Chapter 5, Lesson 3, pages 104–105.

27

Name_____

Range, Median and Mode

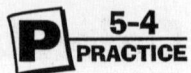

The fourth-grade class at Blue Hill School collects and recycles aluminum cans. The table shows how many cans the students collected in March. Use data from the table for problems 1–3.

1. Find the range, median, and mode from the table.

   Range: _____

   Median: _____

   Mode: _____

Aluminum Cans Collected in March	
Student	Number of Cans
Eddie	24
Maria	32
Emilio	29
Jennifer	26
Mai Ling	28
Frank	29
Tanesha	27

2. What does the mode tell you about this data?

   _____

   _____

3. What does the median tell you about this data?

   _____

   _____

   _____

Complete the table.

Data	Order Data from Least to Greatest	Range	Median	Mode
4. 6, 8, 8, 9, 5, 4, 8, 7, 5				
5. 30, 35, 29, 42, 35, 35, 40				
6. 30, 19, 21, 17, 25, 23, 25				
7. 20, 80, 40, 50, 90, 60, 50				
8. 78, 85, 100, 100, 95, 92, 88				
9. $9, $13, $23, $15, $13				

28 Use with Grade 4, Chapter 5, Lesson 4, pages 106–107.

Name_____

Collect and Organize Data

1. Complete the tally table and line plot for the following data.

Number of Miles Run Each Day by the Members of the Fleet-Footed Club

```
3   2   5   4   6   3   1   5   4   3   2   6
4   3   5   3   2   2   1   5   4   3   6   3
2   5   3   1   4   2   5   6   2   3   2
```

Number of Miles Run Each Day by the Members of the Fleet-Footed Club

Number of Miles	Tally	Total
1	///	
2	//// ////	
3	//// ////	
4	////	
5	//// /	
6	////	

Number of Miles Run Each Day by Members of The Fleet-Footed Club

1   2   3   4   5   6

Use the line plot to answer the questions.

2. How many miles did the greatest number of students run? _____

3. How many members ran 6 miles a day? _____

4. How many members ran 4 miles or more a day? _____

5. How many more members ran 4 miles a day than ran 1 mile a day? _____

6. How many members are in the club? _____

7. Use the data below to make a stem-and-leaf plot on a separate sheet of paper.

Ages of Fleet-Footed Club Members

```
15   17   19   28   31   16   18   22   32   20
20   13   18   25   23   30   27   20   14   21
```

8. What statement can you make about the data in your stem-and-leaf plot?

_____

_____

Use with Grade 4, Chapter 5, Lesson 5, pages 108–110.

Problem Solving: Skill
Identify Extra and Missing Information

Solve. Identify extra or missing information in each problem.

1. A round-trip first-class ticket from St. Louis to San Diego costs $1,600. A round-trip coach ticket costs $359. The Howards buy 3 tickets. How much do they spend?

   _____
   _____

2. A train leaves Rocky Mount, NC, at 1:16 P.M. The train arrives in Petersburg, VA, at 2:45 P.M. and in Richmond, VA, at 3:22 P.M. How long is the trip from Rocky Mount to Richmond?

   _____
   _____

3. A bus leaves the terminal at 6:10 P.M. It makes its first stop at 6:30 P.M. and its second stop at 6:55 P.M. When will the bus arrive at its third stop?

   _____
   _____
   _____

4. Samantha takes a train to New York City. She catches the train at 7:25 A.M. The train stops in Newark at 7:41 A.M. The train arrives in New York at 7:59 A.M. How much time does Samantha's ride take?

   _____
   _____
   _____

Mixed Strategy Review
Solve. Use any strategy.

5. Denzel has 3 rows of shelves in his bedroom. Books, games, or CDs occupy each shelf. The middle shelf holds CDs. If the top shelf does not hold books, which shelf holds games?

   _____

   Strategy: _____

6. Arlene spent $30 for a jacket. She now has $5 left. How much money did Arlene have before she bought the jacket?

   _____

   Strategy: _____

Name _____

Choose the Best Graph

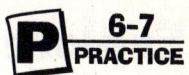

Make an appropriate graph for the data given.

1.

Ms. Adams' Earnings	
Month	Earnings
January	$1,200
February	$1,300
March	$1,000
April	$600
May	$200

2.

City Populations	
City	Population
Danville	8,000
Essex	4,500
Oxford	7,000
Ashton	6,500
Barnard	3,000

3.

Favorite Animals	
Animal	Votes
Cat	10
Dog	5
Horse	3
Snake	2

4.

Test Scores	
Student	Score
Bill	81
Juan	93
Mia	80
Frannie	95

Problem Solving

Solve. Choose the best type of graph for the data. Explain your choice.

5. The number of students in each grade _____
_____

6. The temperature readings over a five-hour period _____
_____

7. The number of votes in an election _____
_____

8. The ages of family members _____
_____

Use with Grade 4, Chapter 6, Lesson 7, pages 132–133.

37

Name_____

Explore the Meaning of Multiplication • Algebra

7-1

Write a multiplication equation for each model.

1. ○ ○ ○ ○ ○
   ○ ○ ○ ○ ○

   _____

2. ○ ○ ○ ○ ○ ○ ○ ○

   _____

3. ○ ○ ○ ○
   ○ ○ ○ ○
   ○ ○ ○ ○

   _____

4. ○ ○ ○ ○ ○ ○ ○ ○
   ○ ○ ○ ○ ○ ○ ○ ○

   _____

5. ○ ○ ○
   ○ ○ ○
   ○ ○ ○
   ○ ○ ○
   ○ ○ ○
   ○ ○ ○

   _____

6. ○ ○ ○ ○ ○ ○
   ○ ○ ○ ○ ○ ○
   ○ ○ ○ ○ ○ ○
   ○ ○ ○ ○ ○ ○
   ○ ○ ○ ○ ○ ○
   ○ ○ ○ ○ ○ ○
   ○ ○ ○ ○ ○ ○

   _____

Find each product. Use models to help.

7.	6	8.	7	9.	3	10.	7	11.	6	12.	7
	×6		×7		×5		×3		×0		×5

13.	5	14.	8	15.	4	16.	9	17.	6	18.	4
	×8		×7		×6		×5		×8		×8

19. 8 × 8 = _____ 20. 2 × 6 = _____ 21. 9 × 6 = _____ 22. 9 × 8 = _____

23. 3 × 3 = _____ 24. 6 × 7 = _____ 25. 2 × 3 = _____ 26. 6 × 9 = _____

27. 8 × 6 = _____ 28. 3 × 6 = _____ 29. 1 × 9 = _____ 30. 9 × 3 = _____

38

Use with Grade 4, Chapter 7, Lesson 1, pages 148–149.

Name _____

Multiply by 5, 7, 8, 9, and 10

7-4 PRACTICE

Multiply.

1. $5 \times 7 =$ ____
2. $9 \times 7 =$ ____
3. $1 \times 8 =$ ____
4. $9 \times 9 =$ ____
5. $3 \times 8 =$ ____
6. $8 \times 7 =$ ____
7. $4 \times 9 =$ ____
8. $2 \times 8 =$ ____
9. $3 \times 7 =$ ____
10. $6 \times 9 =$ ____
11. $7 \times 8 =$ ____
12. $7 \times 7 =$ ____
13. $5 \times 10 =$ ____
14. $2 \times 9 =$ ____
15. $0 \times 7 =$ ____
16. $1 \times 9 =$ ____
17. $6 \times 8 =$ ____
18. $4 \times 7 =$ ____
19. $8 \times 9 =$ ____
20. $4 \times 8 =$ ____

21. 5 × 9
22. 7 × 2
23. 9 × 8
24. 9 × 3
25. 8 × 0
26. 10 × 9

27. 8 × 8
28. 2 × 8
29. 7 × 1
30. 6 × 7
31. 9 × 1
32. 9 × 6

33. 8 × 4
34. 9 × 2
35. 7 × 3
36. 8 × 3
37. 7 × 5
38. 8 × 6

Algebra Find the rule. Then complete the table.

39.

Rule:						
0	1	2	3	4	5	6
0	9	18	27			

40.

Rule:						
0	1	2	3	4	5	6
0	8	16	24			

Problem Solving

Solve.

41. Nathan puts 10 cards on each of 8 pages in an album. How many cards does he put in the album?

42. A marching band has 5 rows with 9 students in each row. How many students are in the marching band?

Use with Grade 4, Chapter 7, Lesson 4, pages 156–158.

Name _____

Problem Solving: Skill
Choose an Operation

Solve. Tell how you chose the operation.

1. Georgia puts coins in an album. There are 8 pages in the album. Each page has slots for 8 coins. How many coins can Georgia put in the album?

   _____
   _____

2. Dina has 37 international dolls. Maxine has 26 international dolls. Who has more dolls? How many more does she have?

   _____
   _____

3. Ben buys 9 packs of dinosaur stickers. There are 6 stickers in each pack. How many stickers does Ben buy?

   _____
   _____

4. Melanie has a collection of 242 stamps. At a stamp convention, she buys 19 more stamps. How many stamps does Melanie have now?

   _____
   _____

Mixed Strategy Review
Solve.

5. James collects model cars. He has 48 model cars. On his birthday, James gets 7 more cars. How many model cars does James have in all?

6. Wendy has 10 flower stickers. She gives away 7 flower stickers. How many flower stickers does Wendy have left?

Name _____

Explore Square Numbers

P 7-6
PRACTICE

Use models to find the square number.

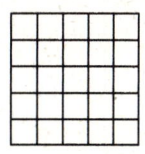

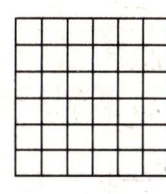

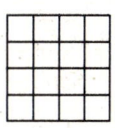

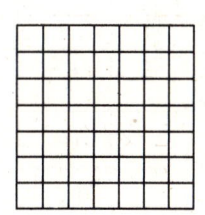

1. 3 × 3 = _____

2. 4 × 4 = _____

3. 2 × 2 = _____

4. 5 × 5 = _____

Find the product. Show your work.

5. 7 × 7 = _____   **6.** 0 × 0 = _____   **7.** 10 × 10 = _____

8. 9 × 9 = _____   **9.** 1 × 1 = _____   **10.** 6 × 6 = _____

11. 5 × 5 = _____   **12.** 4 × 4 = _____   **13.** 8 × 8 = _____

Use with Grade 4, Chapter 7, Lesson 6, pages 162–163.

Name: Ashtin

Multiplication Table and Patterns • Algebra

P 7-7 PRACTICE

Complete the table.

×	0	1	2	3	4	5	6	7	8	9	10	11	12
0	0	0	0	0	0	0	0	0	0	0	0	0	0
1	0	1	2	3	4	5	6	7	8	9	10	11	12
2	0	2	4	6	8	10	12	14	16	18	20	22	24
3	0	3	6	9	12	15	18	21	24	27	30	33	36
4	0	4	8	12	16	20	24	28	32	36	40	44	48
5	0	5	10	15	20	25	30	35	40	45	50	55	60
6	0	6	12	18	24	30	36	42	48	54	60	66	72
7	0	7	14	21	28	35	42	49	56	63	70	77	84
8	0	8	16	24	32	40	48	56	64	72	80	88	96
9	0	9	18	27	36	45	54	63	72	81	90	99	108
10	0	10	20	30	40	50	60	70	80	90	100	110	120
11	0	11	22	33	44	55	66	77	88	99	110	121	132
12	0	12	24	36	48	60	72	84	96	108	120	132	144

Use the table to multiply.

1. $9 \times 8 = 72$
2. $3 \times 12 = 36$
3. $11 \times 11 = 121$
4. $4 \times 12 = 48$

5. 12 × 8 = 96
6. 12 × 12 = 144
7. 12 × 7 = 84
8. 10 × 10 = 100
9. 11 × 7 = 77
10. 12 × 9 = 108

11. What is the pattern of odd and even numbers in the 3 row or 3 column?
even-odd-even-odd-even-odd

12. What is the pattern of odd and even numbers in the 4 row or 4 column?
even-even-even-even

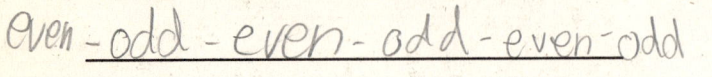

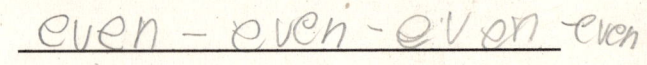

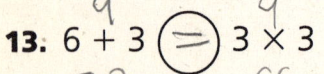

Compare. Write >, <, or =.

13. $6 + 3 \; (=) \; 3 \times 3$ [9 = 9]
14. $15 - 7 \; (<) \; 2 \times 7$ [8, 14]
15. $4 \times 8 \; (<) \; 25 + 4$ [32, 29]
16. $9 \times 7 \; (<) \; 6 \times 11$ [63, 66]
17. $9 + 7 \; (=) \; 4 \times 4$ [16, 16]
18. $12 - 4 \; (>) \; 2 \times 3$ [8, 6]

44

Use with Grade 4, Chapter 7, Lesson 7, pages 164–165.

Name _____

Divide by 2 Through 12

P 8-3
Practice

Divide.

1. 12 ÷ 2 = _____
2. 24 ÷ 3 = _____
3. 32 ÷ 4 = _____
4. 35 ÷ 5 = _____
5. 54 ÷ 6 = _____
6. 56 ÷ 7 = _____
7. 64 ÷ 8 = _____
8. 81 ÷ 9 = _____
9. 40 ÷ 8 = _____
10. 48 ÷ 6 = _____
11. 49 ÷ 7 = _____
12. 27 ÷ 3 = _____
13. 30 ÷ 5 = _____
14. 36 ÷ 4 = _____
15. 72 ÷ 9 = _____
16. 90 ÷ 10 = _____
17. 66 ÷ 11 = _____
18. 96 ÷ 12 = _____

19. 2)18
20. 3)18
21. 4)24
22. 7)14
23. 8)16

24. 7)63
25. 6)42
26. 9)63
27. 5)45
28. 8)72

29. 12)72
30. 11)77
31. 10)80
32. 11)99
33. 12)108

Problem Solving
Solve.

34. There are 84 decorative eggs at a museum. The museum curator places them in display cases. She places 12 eggs in each case. How many cases does she use?

_____

35. Mrs. Pavlik buys a box of 11 Ukrainian eggs to give to one of her grandchildren. She spends $44.00 on the eggs. How much does each egg cost?

_____

36. Marla has 63 foreign stamps in her album. She has 9 stamps on each page. How many pages in Marla's album have stamps?

_____

37. Alan has 25 foreign stamps on 5 pages of his album. Each page has the same number of stamps. How many stamps are on each page?

_____

38. There are 42 tomato plants in rows of 6 plants in each row. How many rows of tomato plants are there?

_____

39. There are 45 tomatoes on 5 tomato plants. Each tomato plant has the same number of tomatoes. How many tomatoes are on each plant?

_____

Use with Grade 4, Chapter 8, Lesson 3, pages 174–176.

Name _____

Missing Factors • Algebra

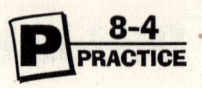

P 8-4 PRACTICE

Complete each fact family.

1. $4 \times 8 = q$ _____
   $8 \times r = 32$ _____
   $32 \div 8 = s$ _____
   $32 \div t = 8$ _____

2. $9 \times 5 = a$ _____
   $5 \times b = 45$ _____
   $45 \div 5 = c$ _____
   $45 \div d = 5$ _____

3. $9 \times 8 = m$ _____
   $8 \times n = 72$ _____
   $72 \div 8 = o$ _____
   $72 \div p = 8$ _____

Find each missing factor.

4. $5 \times k = 30$ _____
   $30 \div 5 = k$ _____

5. $h \times 7 = 56$ _____
   $56 \div 7 = h$ _____

6. $9 \times g = 72$ _____
   $72 \div 9 = g$ _____

7. $9 \times w = 54$ _____
   $54 \div 9 = w$ _____

8. $9 \times y = 63$ _____
   $63 \div 9 = y$ _____

9. $d \times 8 = 48$ _____
   $48 \div 8 = d$ _____

Write a multiplication and division fact family.

10. 8, 5, 40

11. 3, 9, 27

12. 6, 7, 42

13. 9, 8, 72

_____
_____
_____
_____

14. 5, 7, 35

15. 4, 5, 20

16. 6, 9, 54

17. 5, 9, 45

_____
_____
_____
_____

Divide. What patterns do you see?

18. $4 \div 4 = \square$ $8 \div 8 = \square$ $9 \div 9 = \square$ $6 \div 6 = \square$

_____

19. $0 \div 7 = \square$ $0 \div 8 = \square$ $0 \div 1 = \square$ $0 \div 5 = \square$

_____

Use with Grade 4, Chapter 8, Lesson 4, pages 178–181.

Name _____

Problem Solving: Strategy
Work Backward

8-5 PRACTICE

Work backward to solve.

1. Carol had $10 less yesterday than she does today. Yesterday she had $15. How much does Carol have today?

   _____

2. J.R. had 5 baseball cards. Then he bought some more baseball cards at the store. Now J.R. has 9 baseball cards. How many cards did J.R. buy?

   _____

3. Mr. Robinson and Ms. Alvirez drive to the same movie theater. Mr. Robinson drives twice as far as Ms. Alvirez. Ms. Alvirez drives 15 miles. How far does Mr. Robinson drive?

   _____

4. Suki has 4 times as many New York quarters as Georgia quarters. She has 24 New York quarters. How many Georgia quarters does Suki have?

   _____

Mixed Strategy Review
Solve. Use any strategy.

5. Barry makes letters for a sign that reads "Free Field Trip Sign-Up Sheet." Which letter does Mark need to make the most of?

   _____
   Strategy: _____

6. Mr. Carlson has $424. He spends $29 on gasoline. How much money does Mr. Carlson have left?

   _____
   Strategy: _____

7. **Health** Walking a mile burns about 110 calories. About how many calories would you burn if you walked 2 miles?

   _____
   Strategy: _____

8. **Write a problem** that can be solved by working backward. Share it with others.

   _____
   _____

Use with Grade 4, Chapter 8, Lesson 5, pages 182–183.

Name _____

Expressions and Equations • Algebra

P 8-6 PRACTICE

Find the value of each expression.

1. $3 \times (5 - 1)$
2. $(7 + 1) \div 2$
3. $12 - (6 \div 2)$

_____ _____ _____

Circle the best expression.

4. Mark spent 10 minutes a day cleaning his room for 3 days and 15 minutes on the fourth day.

   A. $(10 \times 3) + 15$

   B. $10 \times (3 + 15)$

5. Jennifer had 20 stickers. She bought 10 more stickers. Then she gave half of her stickers to Melanie.

   A. $20 + (10 \div 2)$

   B. $(20 + 10) \div 2$

Solve each equation.

6. $(2 \times 6) + 10 = d$

   $d =$ _____

7. $8 + (5 \times 5) = z$

   $z =$ _____

8. $(14 - 7) \times 3 = n$

   $n =$ _____

Find the value of the expression for the value given.

9. $(x + 2) \times 2$ for $x = 3$
10. $x + (4 \times 5)$ for $x = 10$
11. $8 - (15 \div x) \times 2$ for $x = 5$

_____ _____ _____

Problem Solving

Solve. Use data from the chart for problems 12 and 13.

12. Last week, Karla bought 3 pens and a ruler. How much did she spend?

    _____

13. This week, all items are half price. How much will Karla pay for a ruler and a notebook?

    _____

Item	Cost
pen	$3
ruler	$2
notebook	$4

50

Use with Grade 4, Chapter 8, Lesson 6, pages 184–186.

Name _____

Patterns and Properties • Algebra

9-1 PRACTICE

Complete.

1. $3 \times 2 = a$ $a =$ _____
   $3 \times b = 60$ $b =$ _____
   $c \times 200 = 600$ $c =$ _____
   $3 \times 2{,}000 = d$ $d =$ _____

2. $5 \times 8 = e$ $e =$ _____
   $5 \times c = 400$ $f =$ _____
   $g \times 800 = 4{,}000$ $g =$ _____
   $5 \times 8{,}000 = h$ $h =$ _____

Multiply. Use patterns.

3. 80 × 6
4. 70 × 8
5. 40 × 5
6. 60 × 7
7. 90 × 6

8. 400 × 5
9. 800 × 6
10. 700 × 9
11. 2,000 × 4
12. 3,000 × 6

13. $90 \times 5 =$ _____
14. $4 \times 90 =$ _____
15. $5 \times 600 =$ _____
16. $700 \times 8 =$ _____
17. $9 \times 600 =$ _____
18. $700 \times 4 =$ _____
19. $2{,}000 \times 8 =$ _____
20. $5{,}000 \times 7 =$ _____
21. $8 \times 4{,}000 =$ _____

Find each missing number.

22. $a \times 5 = 300$
    $a =$ _____
23. $b \times 4 = 320$
    $a =$ _____
24. $2 \times c = 180$
    $c =$ _____
25. $3 \times a = 900$
    $a =$ _____
26. $6 \times b = 3{,}600$
    $b =$ _____
27. $c \times 8 = 72{,}000$
    $c =$ _____

Problem Solving

Solve.

28. Stamps are sold in rolls of 100. How many stamps are in 9 rolls?

29. A ream of paper is 500 sheets of paper. How many sheets are in 7 reams?

_____   _____

Use with Grade 4, Chapter 9, Lesson 1, pages 202–203.

Name _____

Explore Multiplying by 1-Digit Numbers

P 9-2 PRACTICE

Use place-value models to multiply.

1. 21
   × 7

2. 38
   × 5

3. 54
   × 2

4. 49
   × 6

5. 17
   × 4

6. 25
   × 9

7. 53
   × 4

8. 28
   × 7

9. 61
   × 8

10. 39
    × 2

11. 62
    × 2

12. 38
    × 4

13. 91
    × 3

14. 46
    × 5

15. 78
    × 6

16. 98
    × 5

17. 76
    × 6

18. 24
    × 9

19. 56
    × 7

20. 48
    × 8

21. 66
    × 6

22. 77
    × 7

23. 94
    × 3

24. 59
    × 4

25. 44
    × 9

26. 24
    × 7

27. 19
    × 8

28. 67
    × 5

29. 84
    × 4

30. 76
    × 7

31. 5 × 26 = _____

32. 37 × 8 = _____

33. 45 × 6 = _____

34. 38 × 4 = _____

35. 7 × 22 = _____

36. 9 × 49 = _____

37. 8 × 67 = _____

38. 35 × 4 = _____

39. 99 × 3 = _____

Use with Grade 4, Chapter 9, Lesson 2, pages 204–205.

Name _____

Multiply by 1-Digit Numbers

P PRACTICE 9-3

Multiply.

1. 73 × 3	**2.** 44 × 5	**3.** 31 × 7	**4.** 68 × 8	**5.** 32 × 9
6. 65 × 5	**7.** 33 × 6	**8.** 96 × 3	**9.** 88 × 4	**10.** 74 × 5
11. 85 × 4	**12.** 77 × 6	**13.** 97 × 2	**14.** 66 × 8	**15.** 94 × 3
16. 44 × 4	**17.** 77 × 7	**18.** 18 × 9	**19.** 38 × 8	**20.** 99 × 6

21. 55 × 5 = _____   **22.** 75 × 6 = _____   **23.** 8 × 47 = _____

24. 6 × 39 = _____   **25.** 2 × 98 = _____   **26.** 84 × 6 = _____

27. 4 × 52 = _____   **28.** 63 × 7 = _____   **29.** 29 × 9 = _____

30. Multiply 63 by 8. _____   **31.** Multiply 78 by 4. _____

32. Multiply 37 by 6. _____   **33.** Multiply 45 by 5. _____

34. Multiply 56 by 7. _____   **35.** Multiply 82 by 3. _____

Problem Solving

Solve.

36. A rectangle is 5 tiles wide by 13 tiles high. How many tiles are in the rectangle?

37. Books are stacked in 3 stacks with 17 books in each stack. How many books are in the stacks?

_____

Use with Grade 4, Chapter 9, Lesson 3, pages 206–208

Name _____

Estimating Products

9-4 PRACTICE

Estimate each product.

1. 5 × 21 = _____
2. 3 × 39 = _____
3. 7 × $46 = _____
4. 85 × 6 = _____
5. 17 × 9 = _____
6. 81 × 3 = _____
7. 2 × $298 = _____
8. 4 × 305 = _____
9. 478 × 6 = _____
10. 5 × 784 = _____
11. 612 × 9 = _____
12. 6 × 556 = _____
13. 2 × 1,987 = _____
14. 3 × $2,126 = _____
15. 7 × 1,905 = _____
16. 8 × 3,495 = _____
17. 4,723 × 4 = _____
18. 5 × $7,118 = _____

19. 41 × 6
20. 28 × 7
21. 96 × 2
22. 17 × 8
23. 31 × 9

24. 255 × 4
25. 488 × 3
26. 563 × 5
27. 2,307 × 5
28. 7,596 × 6

Algebra Compare. Write > or <.

29. 2 × 36 ◯ 1 × 49
30. 96 × 3 ◯ 68 × 4
31. 6 × 28 ◯ 5 × 41
32. 97 × 1 ◯ 89 × 2
33. 6 × 105 ◯ 4 × 209
34. 396 × 4 ◯ 106 × 9
35. 5 × 423 ◯ 6 × 523
36. 3 × 666 ◯ 2 × 366
37. 4 × 712 ◯ 3 × 412

Problem Solving

Solve.

38. The volunteer ambulance group orders 6 first aid kits. Each kit costs $39. About how much does it cost for 6 kits?

39. An ambulance travels about 386 miles a day. About how many miles does it travel in a week?

Name _____

Problem Solving: Skill
Use an Overestimate or Underestimate

P 9-5 PRACTICE

Form a conclusion about whether you would use an overestimate or an underestimate. Then solve each problem.

1. On Wednesday, a group of 98 students will visit the national forest. Each student will get a nature guide fact book. The books come in boxes of 32. The park rangers have 3 boxes of fact books. Are there enough fact books to go around?

    Should you use an overestimate or an underestimate to solve this problem? Explain.

    _____

    _____

    Are there enough fact books so each student can get a book? _____

2. The park charges $16 per day to use a campsite. The Nolans want to use a campsite for 4 nights. They have $80 set aside for using a campsite. Have the Nolans set aside enough money?

    Should you use an overestimate or an underestimate to solve this problem? Explain.

    _____

    _____

    Have the Nolans set aside enough money? _____

3. John, Marla, and Mia each like a different sport, either football, soccer, or basketball. John likes soccer. Mia does not like football. Who likes football?

    _____

4. Caroline had 30 tulip bulbs to plant. She has 12 left. How many bulbs has Caroline already planted?

    _____

Use with Grade 4, Chapter 9, Lesson 5, pages 212–213.

55

Name _____

Multiplying Greater Numbers

P 10-1 PRACTICE

Multiply. Check for reasonableness.

1. 693 × 4
2. 907 × 5
3. 368 × 9
4. $601 × 3

5. 2,901 × 2
6. 1,999 × 7
7. 8,072 × 8
8. $38.88 × 4

9. 6 × 2,369 = _____
10. 7 × 5,786 = _____
11. 3 × 4,964 = _____
12. 9 × $1,288 = _____
13. 5 × 19,091 = _____
14. 8 × 12,967 = _____

15. Multiply 3,687 by 8. _____
16. Multiply 1,096 by 9. _____

Algebra Complete the table.

17.
Input	12	15	18	21	24
Output	48	60			

18.
Input	1	2	3	4	5
Output	37	74			

Problem Solving
Solve.

19. Maria made 9 trips between New York City and Los Angeles. Each trip cost $498. How much did the 9 trips cost?

_____

20. A company buys 8 computers. Each computer costs $2,245. How much does the company spend on the 8 computers?

_____

Name_____

Multiply Using Mental Math

P 10-2 PRACTICE

Multiply mentally.

1. 3 × 203 = _____
2. 210 × 4 = _____
3. 5 × 103 = _____

4. 104 × 6 = _____
5. 7 × 203 = _____
6. 430 × 3 = _____

7. 5 × 204 = _____
8. 501 × 9 = _____
9. 8 × 306 = _____

10. 2,003 × 2 = _____
11. 5 × 1,004 = _____
12. 2,003 × 3 = _____

Find only the products between 2,000 and 35,000.

13. 3 × 903 = _____
14. 9 × 4,006 = _____
15. 8 × 410 = _____

16. 7 × 6,003 = _____
17. 3 × 4,003 = _____
18. 5 × 5,002 = _____

Problem Solving

Solve. Use data from the table for problems 19–20.

19. Mrs. Chan bought 2 T-shirts and 3 pairs of jeans for her children. What was the cost of her purchases?

   _____

20. Mr. Rubens bought 4 sweatshirts and 2 jackets for his children. What was the cost of his purchases?

   _____

Sportswear for Kids	
T-Shirt	$ 9.05
Sweatshirt	$12.10
Jeans	$15.20
Jacket	$21.40

Use with Grade 4, Chapter 10, Lesson 2, pages 222–223.

Name _____

Choose a Computation Method

P 10-3 PRACTICE

Multiply. Tell which method you used.

1. 577
    ×   5

2. 903
    ×   4

3. 241
    ×   3

4. 459
    ×   8

5. 3,006
    ×   7

6. 6,149
    ×   6

7. 5 × 110 = _____

8. 8 × 475 = _____

9. 3 × 1,627 = _____

10. 9 × 1,111 = _____

11. 6 × 4,020 = _____

12. 7 × 5,844 = _____

Problem Solving
Solve.

13. One full shelf in the library holds 250 books. How many books are on 4 full shelves?

14. The library loans 1,855 books per month. How many books does the library loan in 6 months?

15. Twenty-three students in Mr. Sand's class go to the library each week. Each student checks out 4 books. How many books do they check out in 5 weeks?

16. Oxford Elementary School held a book drive. Each of the 784 students brought in 4 books. How many books were collected?

Functions • Algebra

P 10-4 PRACTICE

Complete each function table. Then write an equation.

1. Roger runs 7 miles more each week than another boy.

x	1	2	3	4	5
y	8	9			

_____

2. One plant produces 8 times more peppers than another plant.

r	1	2	3	4	5
s	8	16			

_____

3. One number is 4 less than 3 times another number.

c	4	5	6	7	8
d	8	11			

_____

4. One number is 8 greater than 2 times another number.

m	1	2	3	4	5
n	10	12			

_____

5. Stella works 4 times as many hours as Jana does.

x	0	1	2	3	4
y	0	4			

_____

6. Liz swims 2 more than 2 times as many laps as Sunny does.

b	0	1	2	3	4
a	2	4			

_____

Problem Solving
Solve.

7. Each of 4 people orders a $8.95 lunch. How much do the 4 lunches cost? Write and solve an equation.

_____
_____
_____

8. Ben buys 3 toys that cost $3 each. How much do the toys cost? Write and solve an equation.

_____
_____
_____

Use with Grade 4, Chapter 10, Lesson 4, pages 226–227.

Name _____

Graphing Functions • Algebra

P PRACTICE 10-5

Complete each table. Then graph the function.

1. $b = 2a$

a	0	1	2	3	4
b	0	2			

2. $y = x + 7$

x	0	1	2	3	4
y	7	8			

3. $g = 3f$

f	1	2	3	4	5
g	3				

4. $s = 4r$

r	1	2	3	4	5
s	4				

5. $n = 3m + 3$

m	0	1	2	3	4
n	3				

6. $y = 2x + 2$

x	1	2	3	4	5
y	4				

7. $q = 2p + 1$

p	0	1	2	3	4
q	1				

8. $l = k + 4$

k	0	1	2	3	4
l	4				

Use with Grade 4, Chapter 10, Lesson 5, pages 228–230.

Name _____

Problem Solving: Strategy
Find a Pattern

P 10-6 PRACTICE

Solve.

1. Annie makes an arrangement of chestnuts. She puts 3 chestnuts in the first row, 6 chestnuts in the second row, and 9 chestnuts in the third row. Describe the pattern. How many chestnuts will be in the fourth row?

   _____
   _____

2. In one desert area, the rabbit population is estimated at 25 in one year, 50 the next year, 100 the third year, and 200 the next year. Describe the pattern. Then estimate the rabbit population for the fifth year.

   _____
   _____

3. Rangers examine trees that fell during a storm. The first tree has 3 annual rings. The second tree has 9 rings. The third tree has 27 rings. The fourth tree has 81 rings. If the pattern continues, how many annual rings does the fifth tree have?

   _____

4. Stan counts robins' nests on his block. One year he counts 4 nests. The next year he counts 9 nests. The third year Stan counts 14 nests. The fourth year he counts 19 nests. If the pattern continues, how many nests will he count in the fifth year?

   _____

Mixed Strategy Review
Solve. Use any strategy.

5. Nick took 40 photos of the desert. He has one photo album with 8 pages and another with 12 pages. Nick wants to put the same number of photos on each page. Which album should he use?

   _____
   Strategy: _____

6. **Social Studies** Colorado's state parks cover 347,000 acres. Connecticut's state parks cover 176,000 acres. How many more acres do state parks cover in Colorado than in Connecticut?

   _____
   Strategy: _____

7. **Write a problem** that you would find a pattern to solve. Share it with others.

   _____
   _____
   _____

Use with Grade 4, Chapter 10, Lesson 6, pages 232–233.

Name _____

Patterns of Multiplication • Algebra

P 11-1 PRACTICE

Find each missing number.

1. $6 \times 8 = s$ $s =$ _____
   $60 \times t = 480$ $t =$ _____
   $60 \times 80 = u$ $u =$ _____
   $60 \times 800 = v$ $v =$ _____

2. $w \times 3 = 21$ $w =$ _____
   $70 \times 3 = x$ $x =$ _____
   $y \times 30 = 2{,}100$ $y =$ _____
   $70 \times 300 = z$ $z =$ _____

Multiply. Use mental math.

3. $60 \times 70 =$ _____
4. $20 \times 60 =$ _____
5. $80 \times 800 =$ _____
6. $30 \times 200 =$ _____
7. $50 \times 40 =$ _____
8. $400 \times 30 =$ _____
9. $600 \times 50 =$ _____
10. $90 \times 70 =$ _____
11. $20 \times 4{,}000 =$ _____
12. $9{,}000 \times 30 =$ _____
13. $3{,}000 \times 70 =$ _____
14. $900 \times 60 =$ _____
15. $80 \times 5{,}000 =$ _____
16. $7{,}000 \times 80 =$ _____
17. $40 \times 800 =$ _____
18. $30 \times 6{,}000 =$ _____
19. $20 \times 500 =$ _____
20. $6{,}000 \times 90 =$ _____
21. $700 \times 40 =$ _____
22. $80 \times 2{,}000 =$ _____
23. $50 \times 5{,}000 =$ _____

Algebra Write the number that makes each sentence true.

24. $30 \times j = 9{,}000$ $j =$ _____
25. $s \times 70 = 2{,}800$ $s =$ _____
26. $60 \times b = 24{,}000$ $b =$ _____
27. $400 \times t = 12{,}000$ $t =$ _____
28. $90 \times q = 8{,}100$ $q =$ _____
29. $p \times 600 = 30{,}000$ $p =$ _____
30. $n \times 300 = 6{,}000$ $n =$ _____
31. $r \times 800 = 40{,}000$ $r =$ _____

Problem Solving
Solve.

32. ABC Hardware has 50 cartons of nails. Each carton has 4,000 nails. How many nails does the store have?

    _____

33. Handy Hardware has 500 boxes of hinges. Each box has 90 hinges. How many hinges does the store have?

    _____

Name _____

Multiply by Multiples of 10

P 11-2 PRACTICE

Multiply.

1. 26
    × 40

2. 47
    × 30

3. 91
    × 20

4. 87
    × 10

5. 23
    × 90

6. 17
    × 80

7. 135
    × 50

8. 207
    × 60

9. 399
    × 50

10. 756
    × 30

11. 498
    × 70

12. 1,038
    × 40

13. 2,226
    × 20

14. 3,510
    × 60

15. 5,503
    × 50

16. 2,375
    × 20

17. 4,009
    × 40

18. 2,490
    × 70

19. 6,967
    × 10

20. 9,075
    × 80

21. 51 × 30 = _____

22. 39 × 80 = _____

23. 67 × 20 = _____

24. 325 × 60 = _____

25. 40 × 608 = _____

26. 999 × 10 = _____

27. 712 × 30 = _____

28. 10 × 3,116 = _____

29. 80 × 1,185 = _____

30. 90 × 4,090 = _____

31. 2,111 × 70 = _____

32. 50 × 5,549 = _____

Algebra Find each missing number.

33. 34 × j = 680 j = _____

34. q × 72 = 2,160 q = _____

35. 99 × a = 7,920 a = _____

36. 56 × m = 1,680 m = _____

37. 861 × b = 77,490 b = _____

38. 1,002 × n = 70,140 n = _____

39. s × 2,108 = 63,240 s = _____

40. 898 × c = 53,880 c = _____

Problem Solving
Solve.

41. Classroom chairs cost $39. How much will 30 chairs cost?

42. A computer costs $2,345. How much will 20 computers cost?

_____   _____

Use with Grade 4, Chapter 11, Lesson 2, pages 250–251.   63

Name _____

Explore Multiplying by 2-Digit Numbers

P 11-3 PRACTICE

Use models on graph paper to help you multiply. You may need to tape grids together.

1. 13 × 22 = _____
2. 43 × 15 = _____
3. 17 × 21 = _____
4. 31 × 18 = _____
5. 25 × 24 = _____
6. 20 × 19 = _____

Multiply. Check your answer.

7. 36 × 12
8. 27 × 41
9. 38 × 14
10. 23 × 22
11. 49 × 13

12. 47 × 34
13. 46 × 14
14. 17 × 25
15. 45 × 35
16. 48 × 20

17. 38 × 27
18. 32 × 15
19. 45 × 25
20. 14 × 15
21. 26 × 34

22. 32 × 18
23. 31 × 25
24. 12 × 46
25. 36 × 36
26. 28 × 44

27. 16 × 40
28. 17 × 17
29. 37 × 26
30. 19 × 27
31. 49 × 30

32. 15 × 23 = _____
33. 30 × 13 = _____
34. 14 × 22 = _____

35. 26 × 21 = _____
36. 30 × 24 = _____
37. 42 × 17 = _____

38. 63 × 15 = _____
39. 50 × 23 = _____
40. 13 × 13 = _____

41. 70 × 14 = _____
42. 32 × 20 = _____
43. 25 × 25 = _____

Name _____

Multiply by 2-Digit Numbers

P 11-4 PRACTICE

Find each product.

1. 26
    × 35

2. 73
    × 51

3. 44
    × 87

4. $0.56
    × 83

5. 29
    × 19

6. $46
    × 35

7. 59
    × 47

8. 77
    × 22

9. 55
    × 15

10. 44
    × 46

11. 79
    × 73

12. 94
    × 61

13. $0.63
    × 58

14. 68
    × 24

15. 51
    × 34

16. 18 × 92 = _____
17. 28 × 19 = _____
18. 86 × 43 = _____
19. 74 × 33 = _____
20. 48 × 26 = _____
21. 31 × $0.18 = _____
22. 77 × 94 = _____
23. 88 × 62 = _____
24. 27 × 34 = _____

Algebra Find each product.

25. (30 + 7) × (10 + 8) = n

26. (60 + 4) × (20 + 9) = v

27. (80 + 1) × (40 + 2) = p

28. (50 + 6) × (70 + 3) = r

29. (90 + 5) × (10 + 1) = q

30. (60 + 6) × (50 + 5) = c

31. (20 + 8) × (70 + 7) = s

32. (40 + 3) × (80 + 4) = b

Problem Solving
Solve.

33. A fence has 28 sections with 18 boards in each section. How many boards are in the fence?

34. Horses on a ranch eat 28 bales of hay each day. How many bales do they eat in 31 days?

Use with Grade 4, Chapter 11, Lesson 4, pages 254–256.

Name_____

Problem Solving: Skill
Solve Multistep Problems

11-5 PRACTICE

Solve.

1. The Diving Club offers 4 beginning diving classes each day. Each class has room for 6 people. How many people can take classes in 30 days?

   _____

2. A fishing guide charges $25 per hour. He works 6 hours per day for 5 days. How much money does the guide earn?

   _____

3. During one week, 5 sailboats are rented for a total of 16 hours each. The rental cost is $25 per hour. Altogether, how much is paid for these rentals?

   _____

4. The aquarium charges $12 admission and $6 for a tour. A group of 20 people goes to the aquarium and takes the tour. How much money does the group spend?

   _____

5. Amanda rents a canoe and a life preserver from 2:00 P.M. to 5:00 P.M. A canoe costs $12 per hour. A life preserver costs $2 per hour. How much does Amanda spend?

   _____

6. Jenny rented a rowboat from 10:45 A.M. to 1:00 P.M. After lunch, she rented another rowboat from 1:45 P.M. to 4:45 P.M. For how many minutes did she rent the boat?

   _____

Mixed Strategy Review

Solve. Use any strategy.

7. George had 3 fewer basketball cards yesterday than he does today. Yesterday he had 9 basketball cards. How many basketball cards does George have today?

   _____

   Strategy: _____

8. John collects shells. On the first day he collected 3 shells. On the second day he collected 8 shells, the third day 13 shells, and on the fourth day 18 shells. How many shells will he gather on the eighth day?

   _____

   Strategy: _____

9. Judy, Lakesha, and Tina each like a different color, either red, green, or blue. Judy likes green. Tina does not like blue. What color does Lakesha like?

   _____

   Strategy: _____

10. Mia has saved $7. How much more money does she need to buy a CD for $13?

    _____

    Strategy: _____

66

Use with Grade 4, Chapter 11, Lesson 5, pages 258–259.

Name _____

Estimate Products

P PRACTICE 12-1

Estimate each product.

1. 49 × 59 _____
2. 55 × 65 _____
3. 41 × 52 _____
4. 18 × 29 _____
5. 98 × 402 _____
6. 71 × 874 _____
7. 61 × $216 _____
8. 42 × 605 _____
9. 81 × 350 _____
10. 23 × 999 _____
11. 85 × 1,211 _____
12. 71 × 2,118 _____
13. 19 × 6,302 _____
14. 29 × 7,907 _____

Algebra Compare. Write > or <.

15. 98 × 27 ◯ 3,000
16. 37 × 196 ◯ 8,000
17. 42 × 84 ◯ 3,200
18. 498 × 16 ◯ 100,000
19. 21 × 423 ◯ 8,000
20. 589 × 36 ◯ 24,000
21. 59 × 689 ◯ 42,000
22. 49 × 188 ◯ 10,000
23. 224 × 41 ◯ 8,000
24. 26 × 42 ◯ 34 × 21
25. 15 × 47 ◯ 59 × 68
26. 34 × 82 ◯ 37 × 58

Problem Solving
Solve.

27. The price of a bus ticket is $58. About how much will tickets cost for a group of 62 passengers?

_____

28. An airline ticket costs $375. About how much will tickets cost for a group of 25 people?

_____

Name _____

Multiplying Greater Numbers

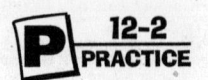

P 12-2 PRACTICE

Multiply. Check that each answer is reasonable

1. 653 × 27
2. 908 × 43
3. 412 × 65
4. 714 × 36

5. 279 × 64
6. 309 × 32
7. $1.26 × 98
8. 305 × 77

9. 4,084 × 43
10. 7,016 × 25
11. 9,148 × 16
12. $50.09 × 31

13. 2,007 × 75
14. $39.85 × 74
15. 6,618 × 91
16. $82.35 × 72

17. 21,107 × 42
18. 46,118 × 27
19. 92,306 × 31
20. $123.95 × 18

21. 53 × 36,219 = _____
22. 26 × $591.05 = _____

23. 36 × 19,962 = _____
24. 71 × 23,401 = _____

25. 42 × $75.53 = _____
26. 83 × 50,276 = _____

Problem Solving
Solve.

27. A box holds 250 table tennis balls. How many table tennis balls can be packaged in 85 boxes?

28. Pencils are packaged with 144 pencils in a box. How many pencils are there in 50 boxes?

68   Use with Grade 4, Chapter 12, Lesson 2, pages 266–268.

Name _____

Problem Solving: Strategy
Make a Graph

P 12-3 PRACTICE

Make a graph for the data in the table. Use data from the graph to solve problems 1 and 2.

Boat Rentals at Lake Willow in July and August	
Type of Boat	**Income from Boat Rentals**
Sailboats	$1,300
Rowboats	$1,100
Paddle boats	$800
Canoes	$1,000

1. Which type of boat generated the most income?

   _____

2. Which type of boat generated the least income?

   _____

3. A beach sells 1,000 passes in 1998, 1,200 passes in 1999, and 1,100 passes in 2000. Suppose you make a pictograph in which each symbol stands for 200 passes. How many symbols would you make for each year?

   _____

4. Suppose you make a graph for the data in problem 3 in which each symbol stands for 100 passes. How many symbols would you make for each year?

   _____

Mixed Strategy Review

Solve. Use any strategy.

5. **Time** Elliot returns from the beach at 4:30 P.M. He spent 2 hours at the beach. It takes 15 minutes for Elliot to travel from his home to the beach. What time did Elliot leave home to go to the beach?

   _____

   Strategy: _____

6. **Create a problem** for which you would make a graph to solve. Share it with others.

   _____
   _____
   _____

Use with Grade 4, Chapter 12, Lesson 3, pages 270–271.

69

Name _____

Multiply Using Mental Math

P PRACTICE 12-4

Multiply. Use mental math.

1. 12 × 30 = _____
2. 40 × 21 = _____
3. 34 × 11 = _____

4. 55 × 18 = _____
5. 60 × 14 = _____
6. 70 × 31 = _____

7. 44 × 22 = _____
8. 80 × 51 = _____
9. 90 × 9 = _____

10. 25 × 50 = _____
11. 30 × 26 = _____
12. 24 × 40 = _____

13. 44 × 15 = _____
14. 52 × 11 = _____
15. 15 × 16 = _____

16. 35 × 22 = _____
17. 61 × 30 = _____
18. 20 × 48 = _____

19. 30 × 19 = _____
20. 65 × 40 = _____
21. 48 × 40 = _____

22. 16 × 21 = _____
23. 25 × 28 = _____
24. 59 × 61 = _____

25. 50 × 14 = _____
26. 35 × 21 = _____
27. 70 × 49 = _____

28. 11 × 62 = _____
29. 90 × 42 = _____
30. 22 × 55 = _____

31. 80
    × 7

32. 41
    × 41

33. 97
    × 11

34. 198
    × 25

35. 38
    × 21

36. 201
    × 11

37. 110
    × 30

38. 55
    × 12

Problem Solving
Solve.

39. Teams of 16 students each are helping clean the park. There are 21 teams in all. How many students in all are helping clean the park?

_____

40. Students are going on a field trip in 20 buses. Each bus carries 35 students. How many students are going on the field trip?

_____

70 Use with Grade 4, Chapter 12, Lesson 4, pages 272–274.

Name _____

Choose a Computation Method

P 12-5 PRACTICE

Multiply. Tell which method you used.

1. 57
   × 15

2. 90
   × 40

3. 41
   × 33

4. 27
   × 81

5. 30
   × 11

6. 64
   × 26

7. 25 × 40 = _____

9. 98 × 47 = _____

8. 83 × 152 = _____

10. 19 × 111 = _____

Algebra Find each missing number.

11. $r × 32 = 832$ _____

12. $28 × n = 1{,}540$ _____

Problem Solving
Solve.

13. The football stadium has 92 rows of seats. There are 25 seats in each row. How many seats are in the football stadium?

14. The baseball stadium has 90 rows of seats. There are 30 seats in each row. How many seats are in the baseball stadium?

Use with Grade 4, Chapter 12, Lesson 5, pages 276–277.

Name _____

Division Patterns • Algebra

13-1 PRACTICE

Complete.

1. 48 ÷ 6 = _____
   480 ÷ 6 = _____
   4,800 ÷ 6 = _____

2. 35 ÷ 5 = _____
   350 ÷ 5 = _____
   3,500 ÷ 5 = _____

3. 16 ÷ 4 = _____
   160 ÷ 4 = _____
   1,600 ÷ 4 = _____

Divide.

4. 3)620
5. 5)250
6. 6)$420
7. 7)560

8. 2)160
9. 3)$270
10. 4)240
11. 8)560

12. 9)7,200
13. 5)3,500
14. 4)2,800
15. 6)$4,200

16. 7)$4,200
17. 9)3,600
18. 3)1,800
19. 2)8,000

20. 120 ÷ 2 = _____
21. $240 ÷ 3 = _____
22. 810 ÷ 9 = _____
23. $450 ÷ 5 = _____
24. 630 ÷ 7 = _____
25. 540 ÷ 9 = _____
26. 3,000 ÷ 6 = _____
27. $7,200 ÷ 8 = _____
28. 4,800 ÷ 8 = _____
29. 3,200 ÷ 8 = _____
30. 5,600 ÷ 7 = _____
31. $3,600 ÷ 4 = _____

Algebra Write the missing number.

32. 200 ÷ _____ = 50
33. 450 ÷ 5 = _____
34. 630 ÷ _____ = 90
35. _____ ÷ 6 = 40
36. 200 ÷ _____ = 40
37. _____ ÷ 8 = 80
38. _____ ÷ 4 = 600
39. 1,500 ÷ _____ = 500
40. 3,000 ÷ 5 = _____

Problem Solving

Solve.

41. There are 150 students in 3 buses. Each bus carries the same number of students. How many students are on each bus?

42. A pet shop has 160 fish in aquariums. Each aquarium has 40 fish. How many aquariums of fish are there?

Name _____

Estimate Quotients

P 13-2 PRACTICE

Estimate. Use compatible numbers.

1. 2)43   2. 4)71   3. 6)521   4. 7)501

5. 3)159   6. 4)171   7. 2)131   8. 9)286

9. 8)650   10. 5)209   11. 9)831   12. 7)2,011

13. 6)3,124   14. 4)3,105

15. 3)5,896   16. 9)46,999

17. 65 ÷ 3   18. 98 ÷ 5   19. 22 ÷ 3
_____   _____   _____

20. 381 ÷ 8   21. 555 ÷ 6   22. 640 ÷ 7
_____   _____   _____

23. 468 ÷ 9   24. 309 ÷ 5   25. 481 ÷ 7
_____   _____   _____

26. 281 ÷ 3   27. 349 ÷ 4   28. 412 ÷ 5
_____   _____   _____

29. 4,124 ÷ 6   30. 1,912 ÷ 9   31. 1,714 ÷ 2
_____   _____   _____

32. 2,186 ÷ 4   33. 2,904 ÷ 7   34. 4,711 ÷ 8
_____   _____   _____

Problem Solving
Solve.

35. Marta travels a total of 850 miles every month to San Francisco on business. If she goes 3 times a month, about how many miles is each round trip?

_____

36. Jeff goes on a 173-mile bike trip. It takes him 9 days from start to finish. About how many miles does he travel each day?

_____

Use with Grade 4, Chapter 13, Lesson 2, pages 294–295.   73

Name _____

Explore Dividing by 1-Digit Numbers

P PRACTICE 13-3

Write a division equation for each model.

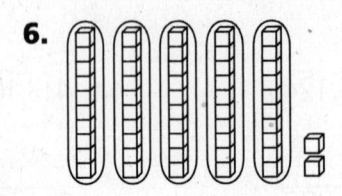

1. _____
2. _____
3. _____

4. _____
5. _____
6. _____

Find each quotient. You may draw place-value models.

7. 6)20   8. 8)29   9. 4)37   10. 9)33

11. 4)51   12. 5)66   13. 6)78   14. 7)83

15. 6)99   16. 7)98   17. 2)55   18. 2)99

19. 41 ÷ 9 = _____   20. 62 ÷ 9 = _____   21. 59 ÷ 7 = _____

22. 88 ÷ 3 = _____   23. 73 ÷ 5 = _____   24. 58 ÷ 4 = _____

25. 67 ÷ 6 = _____   26. 77 ÷ 7 = _____   27. 43 ÷ 2 = _____

Use with Grade 4, Chapter 13, Lesson 3, pages 296–297.

Name_____

Divide by 1-Digit Numbers

P 13-4 PRACTICE

Divide. Check your answer.

1. 2)698
2. 5)$675
3. 3)391
4. 7)785

5. 5)557
6. 8)231
7. 4)$268
8. 8)995

9. 4)398
10. 6)$6.72
11. 3)935
12. 5)457

13. 7)903
14. 2)723
15. 7)836
16. 8)745

17. 9)999
18. 6)377
19. 8)$296
20. 7)779

21. 215 ÷ 3 = _____
22. 367 ÷ 5 = _____
23. 467 ÷ 2 = _____

24. 593 ÷ 4 = _____
25. 298 ÷ 6 = _____
26. 506 ÷ 7 = _____

27. Divide 726 by 7.
28. Divide 834 by 5.
29. Divide 909 by 8.

_____
_____
_____

Algebra Find each missing number.

30. $1{,}065 \div n = 213$
31. $c \div 4 = 168$
32. $690 \div m = 345$

_____
_____
_____

33. $b \div 8 = 116$
34. $585 \div d = 195$
35. $t \div 9 = 111$

_____
_____
_____

36. $(250 + 14) \div x = 44$
37. $(700 + y) \div 7 = 106$
38. $756 \div (r + 3) = 126$

_____
_____
_____

Problem Solving
Solve.

39. Morgan plants 906 pine seedlings. She plants 8 pine seedlings in each row. How many rows are there? How many seedlings are left?

_____

40. The school buys 2,880 tickets to the circus. The tickets are divided equally among 9 classes. How many tickets does each class get?

_____

Use with Grade 4, Chapter 13, Lesson 4, pages 298–300.

Name _____

Problem Solving: Skill
Interpret the Remainder

13-5 PRACTICE

Solve.

1. There are 72 students in the Hockey Club. How many teams of 5 can they make?

   _____

2. The Hockey Club buys 128 ounces of juice. How many 7-ounce cups can they pour?

   _____

3. Paint sets cost $6. The Art Club has $93. If the club buys as many paint sets as it can, how much money will be left over?

   _____

4. There are 132 students at a meeting. The seats are arranged in rows of 8. How many rows of seats are needed?

   _____

5. There are 64 members in the Science Club. They travel to the science fair in cars that can hold 5 members each. How many cars are needed?

   _____

6. There are 83 students. They will sit in rows of 6 seats each. They will start at the front row and fill as many rows as they can. How many students will be in the last row?

   _____

7. Each song played by a DJ is 4 minutes long. How many songs does he play in a music set that is 30 minutes long?

   _____

8. The DJ's assistant distributes neon sunglasses to 50 people at a party. There are 6 glasses in a box. How many boxes should she open?

   _____

Mixed Strategy Review
Solve. Use any strategy.

9. Carla is cooking dinner for friends. She needs 20 minutes to shop for the food, 30 minutes to prepare the food, and 45 minutes to cook the food. If she wants the dinner to be ready at 6:00 P.M., at what time should she start shopping for the food?

   _____

   Strategy: _____

10. Ed, Fred, and Jed each collect a different type of sports card, either baseball cards, football cards, or basketball cards. Ed collects football cards. Fred does not collect baseball cards. What type of card does Jed collect?

    _____

    Strategy: _____

Name _____

Quotients with Zeros

P 13-6 PRACTICE

Divide. Check your answer.

1. 3)620
2. 2)419
3. 9)92
4. 4)839

5. 6)$630
6. 8)$856
7. 7)$763
8. 9)918

9. 5)549
10. 7)748
11. 8)812
12. 2)819

13. 6)620
14. 9)98
15. 3)211
16. 4)827

17. 5)544
18. 8)855
19. 6)657
20. 3)917

21. 2)981
22. 4)835
23. 7)727
24. 8)406

25. 823 ÷ 4 = _____
26. 704 ÷ 5 = _____
27. 981 ÷ 2 = _____

28. 920 ÷ 3 = _____
29. 916 ÷ 7 = _____
30. 845 ÷ 6 = _____

31. 885 ÷ 8 = _____
32. 954 ÷ 5 = _____
33. 965 ÷ 6 = _____

Find only those quotients that are greater than 200.

34. 992 ÷ 3 = _____
35. 920 ÷ 9 = _____

36. 619 ÷ 3 = _____
37. 747 ÷ 4 = _____

38. 818 ÷ 2 = _____
39. 540 ÷ 2 = _____

Problem Solving
Solve.

40. Jenna earns $636 in 6 months by babysitting. If divided evenly, how much is that a month?

41. A family of 4 spends $824 when vacationing. If divided evenly, how much is that per person?

_____

Use with Grade 4, Chapter 13, Lesson 6, pages 304–305.

Name _____

Divide Greater Numbers

P 14-1 PRACTICE

Divide. Check your answer.

1. 5)65,840
2. 4)76,832
3. 2)53,988
4. 6)$90,384

5. 8)33,767
6. 7)45,131
7. 3)6,083
8. 9)27,505

9. 2)14,147
10. 6)31,998
11. 5)23,079
12. 7)65,213

13. $19,328 ÷ 4 = _____
14. 73,895 ÷ 9 = _____

15. 54,620 ÷ 5 = _____
16. 1,841 ÷ 2 = _____

17. 16,697 ÷ 6 = _____
18. 37,986 ÷ 8 = _____

Algebra Find each missing number.

19. $26,480 ÷ n = $5,296
20. 7,240 ÷ v = 1,810
21. 44,356 ÷ r = 11,089

_____   _____   _____

Problem Solving
Solve.

22. The King School holds Junior Olympic games in its sports stadium for 3 days. Each day, every seat in the stadium is full. A total of 17,748 people come to the games. How many seats does the stadium have?

_____

23. The King School raises $75,288 by selling Junior Olympic banners. Each banner costs $6. How many banners does the school sell?

_____

Choose a Computation Method

Divide. Tell which method you used.

1. 5)5,500   2. 9)9,468   3. 3)7,377

Method: _____   _____   _____

4. 9)18,009   5. 4)$20,872   6. 7)2,849

Method: _____   _____   _____

7. 20,328 ÷ 8 = ___   8. 6,713 ÷ 7 = ___   9. 15,369 ÷ 3 = ___

Method: _____   _____   _____

Algebra Find each missing number.

10. 4,050 ÷ s = 810   11. $20,517 ÷ n = $2,931   12. 4,800 ÷ x = 1,200

Problem Solving
Solve.

13. The museum gets 10,500 visitors over 5 days. The same number of people visit the museum each day. How many people visit the museum on each of the 5 days?

14. Suppose the museum makes $17,633 in entrance fees over a period of 7 days. It makes the same amount of money each day. How much does it make in one day?

Name _____

Problem Solving: Strategy
Guess and Check

P 14-3 PRACTICE

Use the guess-and-check strategy to solve.

1. Teri puts 57 dolls in a display case. She puts the same number on each shelf and has 3 dolls left. The case has more than 7 shelves. How many shelves does the case have? How many dolls does each shelf hold?

   _____

2. A group of friends choose cards equally from a deck of 52 cards. There are more than 6 friends. After they have chosen, 4 cards are left. How many friends are there? How many cards does each friend have?

   _____

3. Jamal buys 59 stickers. Stickers come in packs of 5 or 8. How many packs of 5 stickers does Jamal buy? 8 stickers?

   _____

4. There are 36 students in an auditorium. There are twice as many girls as boys. How many girls are there? How many boys are there?

   _____

Mixed Strategy Review
Solve. Use any strategy.

5. Chou makes a display. He puts 1 photo in the first row, 4 photos in the second row, 7 in the third row, and 10 in the fourth row. If the pattern continues, how many photos does Warren put in the fifth row?

   _____

   Strategy: _____

6. **Social Studies** Each of the 50 states in the United States has a state flag. Evelyn wants to make a drawing of each state flag. She has 3 more flags to draw. How many flags has Evelyn drawn?

   _____

   Strategy: _____

7. Sally wants to arrive 20 minutes early for her job. She starts work at 4:15 P.M. It will take her about 20 minutes to walk from school to the job. When should Sally leave?

   _____

   Strategy: _____

8. **Create a problem** that can be solved by using the guess-and-check strategy. Share it with others.

   _____

   _____

   _____

80

Use with Grade 4, Chapter 14, Lesson 3, pages 316–317.

Name _____

Find the Better Buy

P 14-4 PRACTICE

Find each unit price. Find the better buy.

1. 2 ounces for $6.80 _____
   4 ounces for $14.00 _____
   Better buy: _____

2. 3 gallons for $59.91 _____
   5 gallons for $94.90 _____
   Better buy: _____

3. 4 pounds for $10.92 _____
   7 pounds for $19.53 _____
   Better buy: _____

4. 6 pints for $7.14 _____
   9 pints for $14.31 _____
   Better buy: _____

5. 3 yards for $157.44 _____
   4 yards for $199.80 _____
   Better buy: _____

6. 5 inches for $48.40 _____
   9 inches for $78.21 _____
   Better buy: _____

7. 2 quarts for $99.50 _____
   6 quarts for $315.00 _____
   Better buy: _____

8. 4 feet for $2.08 _____
   5 feet for $2.10 _____
   Better buy: _____

Problem Solving

Solve. Use the data from the ad for problems 9–12.

9. What is the unit price for a 2-pound bag of wild bird seed?
   _____

10. What is the unit price for a 5-pound bag of wild bird seed?
    _____

11. What is the unit price of a 9-pound bag of wild bird seed?
    _____

12. Which bag of seed is the best buy?
    _____

Sale on Wild Bird Seed!
2-pound bag for **$3.96**
5-pound bag for **$9.45**
9-pound bag for **$15.75**

Use with Grade 4, Chapter 14, Lesson 4, pages 318–319.

81

Name _____

Explore Finding the Mean

P 14-5 PRACTICE

Use the connecting cubes to find the mean.
Redraw the cubes so that the rows are all the same length.

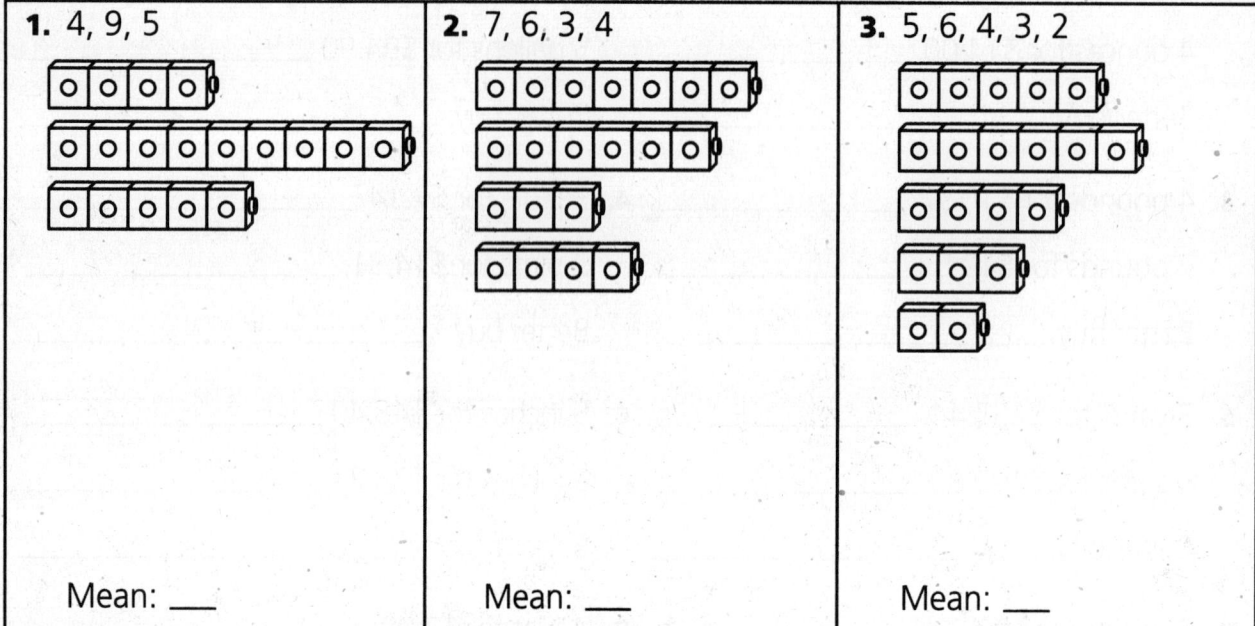

1. 4, 9, 5	2. 7, 6, 3, 4	3. 5, 6, 4, 3, 2
Mean: ___	Mean: ___	Mean: ___

Find the mean. You may use connecting cubes.

4. 2, 2, 9, 9, 8 ___   5. 15, 0, 6 ___   6. 1, 9, 12, 5, 3 ___

7. 5, 10, 15, 20, 0 ___   8. 1, 9, 2, 8, 3, 7 ___   9. 4, 6, 3, 7, 2, 9, 1, 8 ___

10. 10, 10, 30, 30 ___   11. 1, 1, 1, 9, 9, 9, 8, 2 ___   12. 24, 36 ___

13. 20, 15, 20, 25 ___   14. 4, 3, 2, 5, 1, 6, 2, 9 ___   15. 5, 5, 6, 6, 9, 9, 2 ___

16. 5, 10, 15, 20, 30 ___   17. 1, 2, 3, 4, 5, 6, 7, 8, 9 ___   18. 10, 8, 6, 4, 2 ___

Problem Solving
Solve.

19. The students in Homeroom 101 collect soup labels. This week the number of labels brought in to class each day were 8, 6, 10, 6, and 5. What is the mean number of labels brought in each day?

20. Alison plays in a basketball tournament. She scores the following numbers of points in 5 games: 20, 17, 12, 8, and 18. What is her average point total?

_____   _____

82   Use with Grade 4, Chapter 14, Lesson 5, pages 320–321.

Name_____

Find the Mean

P 14-6 PRACTICE

Find the mean.

1. 8, 4, 6, 7, 5 _____

2. 11, 18, 13, 14 _____

3. $25, $48, $77 _____

4. 33, 72, 67, 88 _____

5. $120, $308, $446, $506 _____

6. 823, 665, 482, 619, 781 _____

7. Number of minutes Jason practiced violin this week: 30, 40, 20, 40, 20

8. Number of miles traveled each day: 125, 85, 115, 100, 85, 90

9. Number of rolls of film used each day to take class pictures: 6, 4, 8, 3, 2, 1, 4

10. Number of gallons of gas used each day: 8, 6, 9, 11, 11, 9

11. Number of miles Dorothy ran each day: 6, 8, 7, 9, 10, 11, 12

12. Number of miles a pilot flew each day: 980, 760, 590, 910, 630

13. Number of books Emily read each month: 2, 3, 5, 6, 1, 1

14. Height of six boys in inches: 60, 54, 62, 64, 66, 60

15. Number of bottles of juice on each shelf: 60, 80, 120, 40, 70, 80, 90, 140

16. Number of boxes of cereal eaten by campers each week: 24, 14, 18, 26, 13

Problem Solving
Solve.

17. Kathy trades 42 baseball cards on Monday, 38 on Tuesday, and 40 on Friday. What is the mean number of cards she trades each day?

18. From Thursday through Sunday, Pizza Guy sells 97, 116, 208, and 151 pizzas. What is the average number of pizzas sold each day?

Name _____

Division Patterns • Algebra

P 15-1 PRACTICE

Write the number that makes each equation true.

1. $36 \div 9 = n$ _____
   $360 \div 90 = n$ _____
   $3,600 \div 90 = n$ _____
   $36,000 \div 90 = n$ _____
   $360,000 \div 90 = n$ _____

2. $64 \div 8 = s$ _____
   $640 \div 80 = s$ _____
   $6,400 \div 80 = s$ _____
   $64,000 \div 80 = s$ _____
   $640,000 \div 80 = s$ _____

3. $18 \div b = 6$ _____
   $b \div 30 = 6$ _____
   $1,800 \div 30 = b$ _____
   $18,000 \div 30 = b$ _____
   $180,000 \div 30 = b$ _____

Divide. Use mental math.

4. $60 \overline{)120}$
5. $40 \overline{)2,800}$
6. $70 \overline{)35,000}$
7. $80 \overline{)560,000}$

8. $10 \overline{)\$400}$
9. $70 \overline{)\$21,000}$
10. $40 \overline{)\$2,000}$
11. $90 \overline{)450,000}$

12. $150 \div 30 =$ _____
13. $16,000 \div 80 =$ _____
14. $2,700 \div 90 =$ _____
15. $18,000 \div 20 =$ _____
16. $1,200 \div 20 =$ _____
17. $56,000 \div 70 =$ _____
18. $810 \div 90 =$ _____
19. $42,000 \div 70 =$ _____
20. $3,600 \div 40 =$ _____
21. $45,000 \div 50 =$ _____

Algebra Find each missing number.

22. $140 \div a = 2$
    _____
23. $d \div 70 = 7$
    _____
24. $3,000 \div 60 = x$
    _____

25. $t \div 60 = 70$
    _____
26. $28,000 \div b = 400$
    _____
27. $40,000 \div 50 = y$
    _____

Problem Solving
Solve.

28. A box of 400 stickers is divided equally among 80 students. How many stickers did each student receive?

    _____

29. If 6,300 books are divided equally among 90 libraries, how many books will each library get?

    _____

Name_____

Estimating Quotients

P 15-2 PRACTICE

Estimate the quotient. Choose compatible numbers.

1. 19)389

2. 17)211

3. 18)586

4. 16)789

5. 49)1,585

6. 72)6,280

7. 32)8,920

8. 61)3,256

9. 68)34,912

10. 2,806 ÷ 38

11. 7,903 ÷ 86

12. 1,113 ÷ 31

13. 7,160 ÷ 93

14. 2,806 ÷ 56

15. 2,210 ÷ 48

16. 21)1,732

17. 63)546

18. 53)2,612

19. 41)1,512

20. 78)4,106

21. 86)1,709

Algebra Compare. Write > or <.

22. 396 ÷ 21 ◯ 914 ÷ 31

23. 492 ÷ 68 ◯ 556 ÷ 71

24. 1,947 ÷ 38 ◯ 2,011 ÷ 48

25. 1,300 ÷ 21 ◯ 2,300 ÷ 13

26. 5,106 ÷ 82 ◯ 6,206 ÷ 91

27. 3,100 ÷ 82 ◯ 4,700 ÷ 71

Problem Solving
Solve.

28. Karen drove 283 miles at a speed of 46 miles per hour. About how many hours did she drive?

29. A jet flew 3,116 miles in 6 hours. About how many miles per hour did it fly?

Use with Grade 4, Chapter 15, Lesson 2, pages 342–343.

Name _____

Divide 2-Digit Numbers by Multiples of 10   **P** 15-3 PRACTICE

Divide.

1. 82 ÷ 20 = _____
2. 75 ÷ 10 = _____
3. 51 ÷ 20 = _____

4. 94 ÷ 30 = _____
5. 88 ÷ 20 = _____
6. 87 ÷ 10 = _____

7. 93 ÷ 40 = _____
8. 71 ÷ 30 = _____
9. 97 ÷ 20 = _____

10. 74 ÷ 20 = _____
11. 52 ÷ 10 = _____
12. 67 ÷ 30 = _____

13. 91 ÷ 10 = _____
14. 62 ÷ 40 = _____
15. 94 ÷ 40 = _____

16. 20)61
17. 50)78
18. 40)81
19. 30)63

20. 10)76
21. 20)95
22. 60)84
23. 40)49

24. 10)96
25. 30)59
26. 20)44
27. 50)59

Algebra Find each missing number.

28. 27 ÷ m = 2 R7 _____
29. 51 ÷ k = 1 R21 _____

30. 63 ÷ a = 1 R13 _____
31. 74 ÷ p = 3 R14 _____

32. 71 ÷ y = 3 R11 _____
33. 90 ÷ r = 2 R10 _____

Problem Solving
Solve.

34. Sam puts 76 pencils into packages. Each package has 10 pencils. How many packages does Sam make? How many pencils are left over?

_____

35. Kenya puts 84 cans of tennis balls in boxes. Each box has 20 cans. How many boxes does Kenya fill? How many cans does she have left over?

_____

Use with Grade 4, Chapter 15, Lesson 3, pages 344–345.

Name _____

Explore Dividing by 2-Digit Numbers

P 15-4 PRACTICE

Divide.

1.

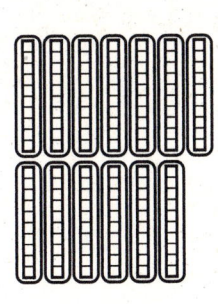

130 ÷ 10 = _____

2.

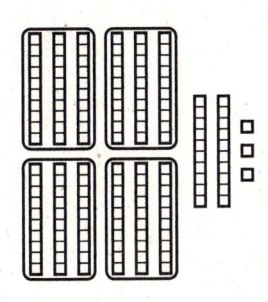

143 ÷ 30 = _____

3.

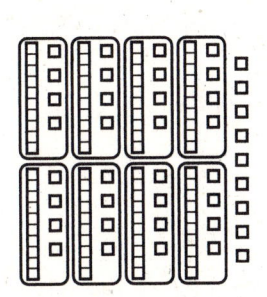

121 ÷ 14 = _____

4.

156 ÷ 18 = _____

Divide. You may use place-value models.

5. 13)87   6. 15)137   7. 12)93   8. 14)125

9. 16)293   10. 17)235   11. 19)258   12. 25)441

13. 135 ÷ 16 = _____   14. 134 ÷ 14 = _____   15. 115 ÷ 15 = _____

16. 282 ÷ 18 = _____   17. 230 ÷ 19 = _____   18. 269 ÷ 24 = _____

Problem Solving
Solve.

19. The dividend is 280. The divisor is 23. What are the quotient and remainder?

_____

20. The dividend is 160. The divisor is 12. What are the quotient and remainder?

_____

Use with Grade 4, Chapter 15, Lesson 4, pages 346–347.

87

Divide by 2-Digit Numbers

15-5 PRACTICE

Divide. Check your answer.

1. 22)952
2. 31)784
3. 66)$7.26
4. 54)760

5. 81)891
6. 29)496
7. 44)530
8. 75)984

9. 26)1,954
10. 17)$11.39
11. 39)2,381
12. 46)3,818

13. 93)8,929
14. 51)3,621
15. 62)$55.18
16. 88)6,518

17. 895 ÷ 24 = _____
18. 907 ÷ 31 = _____
19. 367 ÷ 14 = _____
20. $7.08 ÷ 59 = _____
21. 814 ÷ 36 = _____
22. 531 ÷ 45 = _____
23. 1,467 ÷ 24 = _____
24. $37.76 ÷ 64 = _____
25. 4,780 ÷ 77 = _____
26. $48.59 ÷ 43 = _____
27. 7,900 ÷ 84 = _____
28. 8,930 ÷ 92 = _____

Algebra Solve.

29. (1,700 + 53) ÷ 37 = w _____
30. (1,000 + 160) ÷ 46 = d _____
31. (1,900 + 100) ÷ 29 = v _____
32. (1,600 + 240) ÷ 83 = x _____
33. (2,300 + 70) ÷ (12 × 4) = n _____
34. (1,500 + 80) ÷ (11 × 5) = c _____

Problem Solving

Solve.

35. Mrs. Tallo's class makes 234 ribbons for the Sports Fair. Each student makes the same number of ribbons. There are 18 students in the class. How many ribbons does each student make?

36. Mr. Willow's class of 25 students sells 200 tickets to the Sports Fair. Each student sells the same number of tickets. How many tickets does each student sell?

88

Use with Grade 4, Chapter 15, Lesson 5, pages 348–350.

Name _____

Problem Solving: Skill
Use an Overestimate or Underestimate

15-6 PRACTICE

Solve.

1. Travis makes first-place ribbons for Sports Day. He uses 111 inches of ribbon. Each ribbon is 8 inches long. Underestimate the number of ribbons he makes.

   _____

2. The soccer club makes 100 cups of fruit drink. There are 46 students in the soccer club. Is there enough fruit drink for each student to have 2 cups? Explain.

   _____

3. There are 152 people at the Sports Night Dinner. There are 33 tables. What is the greatest number of people that can sit at a table? Explain.

   _____

4. Mark wants to buy baseball shirts of different teams. Each shirt costs $18. Mark has $62. How many shirts can he buy? Explain.

   _____

5. A pack of 3 pennants costs $8. Maryanne has $30. Is this enough to buy 4 packs of pennants? Explain.

   _____

6. A box of gold medals costs $56. The Sports Committee has $185 to spend on medals. How many boxes can the committee buy? Explain.

   _____

Mixed Strategy Review
Solve. Use any strategy.

7. Jamal has 288 stickers. He has twice as many animal stickers as sports stickers. How many of each kind does Jamal have?

   _____

   Strategy: _____

8. Jamal puts 288 stickers into 24 envelopes. He puts the same number of stickers in each envelope. How many stickers are in each envelope?

   _____

   Strategy: _____

Use with Grade 4, Chapter 15, Lesson 6, pages 352–353.

Name_____

Adjusting Quotients

P 16-1 PRACTICE

Divide. Check your answers.

1. 34)249   2. 26)189   3. 56)469   4. 41)367

5. 51)146   6. 84)626   7. 79)350   8. 63)238

9. 92)810   10. 75)295   11. 39)230   12. 25)186

13. 56)476   14. 69)507   15. 92)546   16. 88)339

17. 44)371   18. 24)129   19. 65)247   20. 57)284

21. 81)482   22. 22)186   23. 45)395   24. 36)299

Algebra Divide only those with quotients between $5.00 and $8.00.

25. 18)$94.50   26. 16)$98.40   27. 14)$60.90   28. 25)$93.75

29. 13)$92.95   30. 11)$99.11   31. 15)$56.25   32. 12)$93.12

Problem Solving
Solve.

33. Candy wants to walk 220 miles in 30 days. If she walks 7 miles every day, will she meet her goal?

_____

34. Jason wants to save $180 in 12 months. How much should he save each month?

_____

90

Use with Grade 4, Chapter 16, Lesson 1, pages 358–360.

Name _____

Choose a Computation Method

P 16-2 PRACTICE

Divide. Tell which method you used.

1. 25)25,250

   Method: _____

2. 49)18,032

   Method: _____

3. 30)6,090

   Method: _____

4. 18)37,818

   Method: _____

5. 74)$4,366

   Method: _____

6. 51)5,610

   Method: _____

7. 9,288 ÷ 36 = _____

   Method: _____

8. 1,250 ÷ 50 = _____

   Method: _____

9. 16,967 ÷ 19 = _____

   Method: _____

Algebra Find each missing number.

10. 40,000 ÷ x = 2,000

    _____

11. $90,090 ÷ n = $3003

    _____

12. 31,125 ÷ y = 1,245

    _____

13. $18,006 ÷ q = $1,500.50

    _____

14. 25,000 ÷ p = 200

    _____

15. 73,928 ÷ r = 9,241

    _____

Problem Solving
Solve.

16. Jenny received 1,872 e-mails last year. What was the average number of e-mails she received each month?

    _____

17. The Publishing Company bought 28 identical computers for its new office. If the total cost of the computers was $56,280, how much did each computer cost?

    _____

Use with Grade 4, Chapter 16, Lesson 2, pages 362–363.

Name _____

Problem Solving: Strategy
Choose a Strategy

P 16-3 PRACTICE

Choose a strategy. Use it to solve the problem.

1. The Sports Committee buys 30 yards of material. The material is cut into banners that are 5 feet long. How many banners are made?

   _____

2. The Sand Trap Golf Shop has 132 golf balls in stock. The golf balls are packed in tubes of 6. How many tubes of golf balls does the store have?

   _____

3. Liam is building a fence around his backyard. The backyard is 24 feet wide and 60 feet long. If Liam uses sections of fencing that are 12 feet long, how many sections does he use?

   _____

4. There are 115 students going to the basketball tournament. Each bus can carry 26 students. How many buses are needed?

   _____

Mixed Strategy Review
Solve. Use any strategy.

5. Tina makes a display of 36 autographed baseballs. She puts 12 baseballs in a large display case. Tina also has 4 smaller display cases. How can she arrange the baseballs in the smaller cases so that each smaller case has an equal number of baseballs?

   _____

   Strategy: _____

6. Francine uses a pattern to make a window display for a sneaker store. The first row has 2 sneakers, the second row has 6 sneakers, the third row has 10, and the fourth row has 14. How many sneakers are in the fifth row?

   _____

   Strategy: _____

7. The Stadium Store sells 450 team photos and 369 individual photos. How many photos does it sell in all?

   _____

   Strategy: _____

8. **Write a problem** that you could solve by drawing a diagram or by writing a division sentence. Share it with others.

   _____

   _____

92

Use with Grade 4, Chapter 16, Lesson 3, pages 364–365.

Name _____

Order of Operations • Algebra

16-4 PRACTICE

Write which operation should be done first.

1. $2 \times 8 + 7$

2. $2 + 3 \times 9$

3. $4 + 10 \div 2$

4. $9 - 2 + 3$

5. $(3 + 2) \times 9$

6. $8 \div (2 + 2)$

7. $6 \div 2 - 1$

8. $1 + 3 \times 5$

9. $10 \div 5 \times 2$

10. $7 - 8 \div 2$

11. $(12 - 4) \div 2$

12. $9 + 2 - 6$

Simplify. Use the proper order of operations.

13. $3 + 2 \times 7 =$ _____

14. $10 \div 2 - 1 =$ _____

15. $9 - 6 \div 2 =$ _____

16. $24 \div 2 - 8 =$ _____

17. $(2 + 6) \times 7 =$ _____

18. $12 - 12 \div 3 =$ _____

19. $(4 + 6) \div 5 =$ _____

20. $12 - 3 + 9 =$ _____

21. $20 \div 5 \div 2 =$ _____

22. $18 \div 9 \times 6 =$ _____

23. $2 \times 8 \div 4 =$ _____

24. $20 - 5 \times 4 =$ _____

25. $2 \times 6 + 4 \times 3 =$ _____

26. $20 \div 2 \times 3 - 6 =$ _____

27. $(2 + 9) \times (7 - 3) =$ _____

28. $4 + (14 - 6) \times 2 + 5 =$ _____

29. $2 \times 9 + 10 \div 5 \times (3 + 2) =$ _____

Problem Solving

Solve.

30. Tamara buys 6 apples for $0.40 each. She has a $0.50-off coupon. How much does Tamara spend? Write an expression and simplify.

31. Steven has 126 photos in an album. He adds 18 more photos to the album. Each page holds 12 photos. Write an expression and simplify to find out how many pages Steven fills.

Use with Grade 4, Chapter 16, Lesson 4, pages 366–368.

Name _____

Explore Nonstandard Units for Length, Width, and Height

P 17-1
PRACTICE

Use nonstandard units to measure. Tell what units you use. Explain why you chose the units you did.

1. the length of your pencil
   _____
   _____

2. the length of your shoe
   _____
   _____

3. the width of your classroom
   _____
   _____

4. the width of your desk
   _____
   _____

5. the height of your teacher
   _____
   _____

6. the height of another fourth-grader
   _____
   _____

Name _____

Explore Customary Length to $\frac{1}{4}$ Inch

P 17-2 PRACTICE

Estimate and then measure. Tell what unit and tool you use.

1. length of a pencil _____

2. height of a desk _____

3. width of a desk _____

4. height of a door _____

5. width of a window _____

6. width of the classroom _____

7. length of a book _____

8. distance you go in a stride _____

9. length of your forearm _____

Measure each line to the closest quarter inch. Write your answer on the line.

10. _____   11. _____

12. _____   13. _____

14. _____   15. _____

16. _____   17. _____

18. _____

19. _____

Use with Grade 4, Chapter 17, Lesson 2, pages 386–387.

95

Name _____

Customary Capacity and Weight

P 17-3 PRACTICE

Estimate and then measure the capacity of each object.

1. a water glass _____
2. a large pot _____
3. a cereal bowl _____
4. a milk carton _____
5. Order the objects above from least to greatest capacity.

_____

Estimate and then measure the weight of each object or objects.

6. an apple _____
7. four potatoes _____
8. two envelopes _____
9. a pencil _____
10. Order the objects above from least to greatest weight.

_____

Circle the letter of the correct estimate.

11. 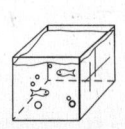 **A.** 5 c **B.** 5 pt **C.** 5 gal

12. 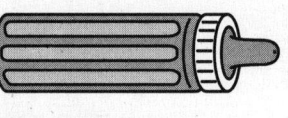 **A.** 1 c **B.** 1 pt **C.** 1 qt

13. 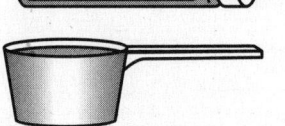 **A.** 6 c **B.** 6 qt **C.** 6 gal

14. **A.** 2 fl oz **B.** 2 c **C.** 2 pt

15. 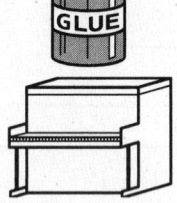 **A.** 500 oz **B.** 500 lb **C.** 500 T

Problem Solving
Solve.

16. A box of Krispy Krunch cereal holds 20 oz. Kyle pours 3 oz of cereal into his bowl. How much cereal is left in the box?

_____

17. Sarah buys a 48-fl-oz bottle of apple juice. How many cups of juice can she pour from the bottle?

_____

Name _____

Converting Customary Units • Algebra

P 17-4 PRACTICE

Write the number that makes each equation true.

1. 7 ft = _____ in.
2. 21 ft = _____ yd
3. 2 mi = _____ yd
4. 60 in. = _____ ft
5. 13 yd = _____ ft
6. 2 mi = _____ ft
7. 8 qt = _____ gal
8. 144 in. = _____ ft
9. 3 pt = _____ c
10. 36 ft = _____ yd
11. 4 ft = _____ in.
12. 12 ft = _____ yd
13. 12 pt = _____ qt
14. 2 lb = _____ oz
15. 48 oz = _____ lb
16. 3 T = _____ lb
17. 10,000 lb = _____ T
18. 2 c = _____ fl oz
19. 3 gal = _____ qt
20. 2 qt = _____ pt
21. 10 c = _____ pt
22. 1 lb 10 oz = _____ oz
23. 1 gal 2 pt = _____ pt
24. 10 ft = ___ yd ___ ft
25. 4 T 800 lb = _____ lb
26. 5 ft 8 in. = _____ in.
27. 13 qt = ___ gal ___ qt

Complete the table.

28.
Gallons	1			
Quarts		12		
Pints			16	
Cups				64

29.
Yards	1		
Feet		9	
Inches			72

30.
Ounces	Pounds
	$\frac{1}{2}$
	$\frac{3}{4}$
16	
32	
48	

31.
Tons	1		
Pounds		6,000	

Problem Solving
Solve.

32. Amy cuts a piece of ribbon 60 in. long. How many feet long is the piece of ribbon?

_____

33. The 6 members of the Brown family drink a total of 3 gallons of milk each week. How much is that per person?

_____

Use with Grade 4, Chapter 17, Lesson 4, pages 392–394.

Name _____

Problem Solving: Skill
Check for Reasonableness

P 17-5 PRACTICE

Solve. Explain your answer.

1. Tyler walks 4 miles from his home to the movie theater. He says he walks more than 20,000 feet. Is his statement reasonable?

   _____
   _____

2. A movie star is 6 feet tall. Meg says that the movie star is more than 80 inches tall. Is her statement reasonable?

   _____
   _____

3. Tammy's sled is 65 inches long. She says the sled is more than 5 feet long. Is her statement reasonable?

   _____
   _____

4. Earl's house is 1,200 yards from the bus stop. Earl says that is 3,600 feet. Is his statement reasonable?

   _____
   _____

5. The popcorn stand sells 100 ounces of popcorn. Ben says this is 1,600 pounds of popcorn. Is his statement reasonable?

   _____
   _____
   _____

6. The refreshment stand sells 144 ounces of peanuts. The manager says that this is more than 10 pounds of peanuts. Is his statement reasonable?

   _____
   _____
   _____

Mixed Strategy Review
Solve. Use any strategy.

7. Darryl runs every day. On the first day he ran 1 mile. On the second he ran 3 miles, on the third he ran 5 miles, and on the fourth he ran 7 miles. How many miles will Darryl probably run on the tenth day?

   _____

   Strategy: _____

8. Ashley and Fiona swim laps in the pool. Ashley swims twice as many laps as Fiona. Fiona swims 10 laps. How many laps does Ashley swim?

   _____

   Strategy: _____

Use with Grade 4, Chapter 17, Lesson 5, pages 396–397.

Name _____

Explore Metric Length

P 18-1 PRACTICE

Estimate and then measure. Tell what unit and tool you use.

1. the width of your classroom _____

2. the largest step you can take _____

3. the width of a window in your classroom _____

4. the distance from the tip of your hand to your elbow _____

5. the thickness of a nickel _____

6. the width of a dime _____

7. your height _____

8. length of a sheet of notebook paper _____

Measure each line to the closest centimeter. Write your answer on the line.

9. _____ 10. _____

11. _____ 12. _____

13. _____ 14. _____

15. _____ 16. _____

17. _____

18. _____

Use with Grade 4, Chapter 18, Lesson 1, pages 402–403.

99

Name _____

Metric Capacity and Mass

P 18-2 PRACTICE

Estimate and then measure the capacity of each object.

1. a water glass _____   2. a large pot _____

3. a cereal bowl _____   4. a milk carton _____

5. Order the objects above from least to greatest capacity.

_____

Estimate and then measure the mass of each object.

6. a box of crayons _____   7. a book _____

8. a paper clip _____   9. a pencil _____

10. Order the objects above from least to greatest mass.

_____

Circle the letter of the best estimate.

11. A. 15 mL   B. 15 L   C. 2 L

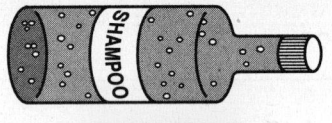

12. A. 3 mL   B. 31 L   C. 310 mL

13. A. 200 mL   B. 200 L   C. 2 mL

Algebra Complete the table.

14.
Liters	1	2	3		
Milliliters	1,000			4,000	

Problem Solving
Solve.

15. Sally buys 1 kg of grapes. She packs 200 g of grapes in her lunch. How many grams of grapes are left?

16. Jim buys 1 L of milk. He drinks 300 mL for breakfast. How many milliliters of milk are left?

_____

100   Use with Grade 4, Chapter 18, Lesson 2, pages 404–406.

Name _____

Convert Metric Units • Algebra

P 18-3 PRACTICE

Write the number that makes each sentence true.

1. 5 m = _____ cm
2. 2 L = _____ mL
3. 7 kg = _____ g
4. 10 mm = _____ cm
5. 5 kg = _____ g
6. 2 m = _____ dm
7. 3,000 mL = _____ L
8. 300 cm = _____ m
9. 4,000 g = _____ kg
10. 6,000 mL = _____ L
11. 40 kg = _____ g
12. 40 cm = _____ dm
13. 700 cm = _____ m
14. 10 L = _____ mL
15. 2 km = _____ m
16. 10,000 g = _____ kg
17. 6,000 cm = _____ m
18. 4 m = _____ mm
19. 20 cm = _____ mm
20. 3 dm = _____ mm
21. 5 L = _____ mL
22. 10 m = _____ cm
23. 5 cm = _____ mm
24. 600 mm = _____ cm
25. 8,000 mm = _____ cm
26. 4,000 m = _____ km
27. 7,000 mL = _____ L
28. 20,000 mL = _____ L
29. 70,000 g = _____ kg

Compare. Write >, <, or =.

30. 5,000 g ◯ 5 kg
31. 20 L ◯ 200 mL
32. 50 cm ◯ 6 dm
33. 60 cm ◯ 6 m
34. 300 cm ◯ 3 m
35. 2,500 mL ◯ 2 L
36. 3 km ◯ 300 m
37. 900 mm ◯ 9 cm
38. 13 L ◯ 1,300 mL
39. 500 dm ◯ 5 dm
40. 7 dm ◯ 7,000 mm
41. 18,000 mL ◯ 18 L

Problem Solving
Solve.

42. Dottie has 1 kg 200 g of food for her cat. How many grams of cat food does she have?

43. A 1-L bottle of water is half full. How many milliliters of water are in the bottle?

_____

_____

Use with Grade 4, Chapter 18, Lesson 3, pages 408–410.

Name _____

Problem Solving: Strategy
Logical Reasoning

P 18-4 PRACTICE

Use logical reasoning to solve each problem.

1. An aquarium worker needs to fill a tank with 10 gallons of water. He has an 8-gallon pail and a 6-gallon pail. How can he use the pails to get exactly 10 gallons of water in the tank?

2. Simon needs to put 9 cups of sea salt into a saltwater tank. He has a 10-cup container and a 7-cup container. How can Simon use the containers to measure 9 cups?

3. The parrot house has 2 times as many birds as the toucan house. The toucan house has 3 more birds than the jay house. The jay house has 6 birds. How many birds do the other houses have?

4. The parrots get food 20 minutes before the toucans. The toucans get food 15 minutes after the jays. The jays get food 30 minutes after Bird World opens. Bird World opens at 10:00 A.M. When does each kind of bird get food?

Mixed Strategy Review
Solve. Use any strategy.

5. **Language Arts** Kenny writes a 740-word review of a play. The review needs to be cut so that it has 500 words. How many words have to be cut?

Strategy: _____

6. There are 24 cars in the theater parking lot. There are 3 times as many 4-door cars as 2-door cars. How many of each kind of car are there?

Strategy: _____

7. A bandstand is 40 feet wide by 80 feet long. It is built from wood planks that are 5 feet wide by 10 feet long. How many planks wide is the bandstand? How many planks long?

Strategy: _____

8. **Write a problem** which you could solve by using logical reasoning. Share it with others.

102

Use with Grade 4, Chapter 18, Lesson 4, pages 412-413.

Name _____

Problem Solving: Skill
Use a Diagram

P 19-6
PRACTICE

Solve.

1. This figure is a parallelogram. Suppose you draw a line segment from point A to point C. The length of this segment is 5 cm. How would you describe the two new figures you made?

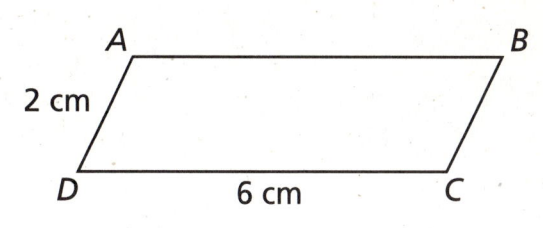

   _____

   _____

Solve. Use data from the illustration to answer problems 2–6.

2. Orson designed this picture frame. What shapes make up the frame? What shape is made by the outer edge of the frame?

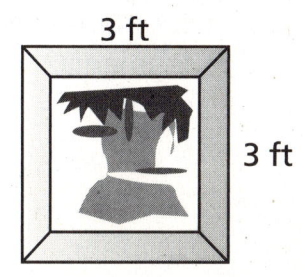

   _____

3. Suppose Robert added 2 feet to the height of the frame, but kept the width the same. What shape would be made by the outer edge of the frame?

   _____

4. Wendy drew a triangle in which three angles were less than 90°. What kind of triangle did she draw?

   _____

5. Robert drew a square. Then he divided the shape into two parts by drawing a line from one corner of the square, through the center, to the opposite corner. Name two ways to describe the two smaller shapes he created.

   _____

6. Max draws a rectangle with sides of 6 inches and 9 inches. He uses one of the short sides of the rectangle as a side of a scalene triangle. Can the lengths of the other two sides of the triangle be 6 inches? Explain.

   _____

   _____

Use with Grade 4, Chapter 19, Lesson 6, pages 448–449.

109

Name_____

Parts of a Circle

PRACTICE 19-7

Identify the parts of a circle.

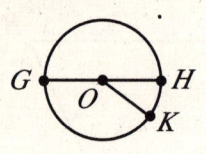

1. (circle with G, O, H, K)

_____

_____

2. (circle with L, M, N)

_____

_____

3. (circle with T, S, V)

_____

_____

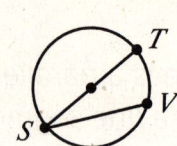

Use data from the circle for problems 4–9. Locate each pair of points on the circle. Name the line segments they create and classify them as parts of a circle.

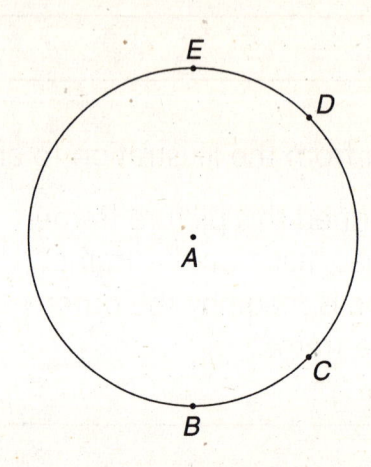

(circle with points E, D, A (center), C, B)

4. A, D

_____

5. E, B

_____

6. C, D

_____

7. D, B

_____

8. A, C

_____

9. B, C

_____

Problem Solving

Solve.

10. Alan drew a chord through the center of a circle. What part of the circle did Alan draw?

_____

11. Make a point below. Use a Triman compass to draw a circle. Then draw and label a diameter, a radius, and a chord.

110

Use with Grade 4, Chapter 19, Lesson 7, pages 450–451.

Name_____

Explore Volume • Algebra

P 20-7 PRACTICE

Find the volume of each rectangular prism.

1.

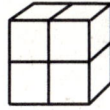

2.

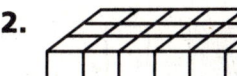

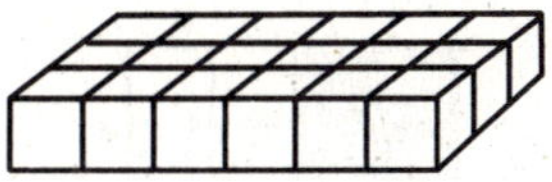

3.

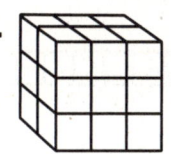

_____ _____ _____

4.

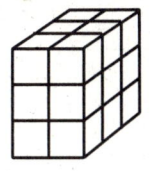

5.

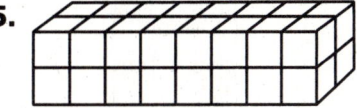

6.

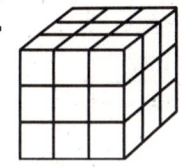

_____ _____ _____

7.

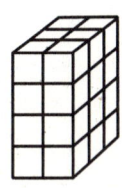

8.

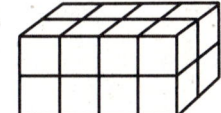

9.

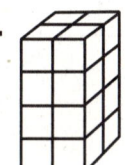

_____ _____ _____

10.

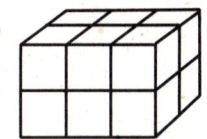

11.

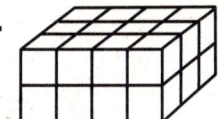

12.

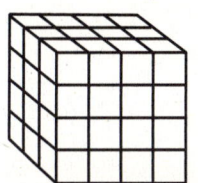

_____ _____ _____

13. length: 9 in.
    width: 5 in.
    height: 4 in.

14. length: 5 m
    width: 8 m
    height: 6 m

15. length: 7 cm
    width: 2 cm
    height: 8 cm

16. length: 10 ft
    width: 12 ft
    height: 5 ft

_____ _____ _____ _____

Use with Grade 4, Chapter 20, Lesson 7, pages 478–479.

Name _____

Parts of a Whole

P **21-1**
PRACTICE

Write a fraction for the part that is shaded.

1.

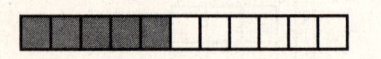

2.

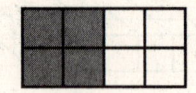

3.

_____   _____   _____

4.

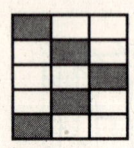

5.

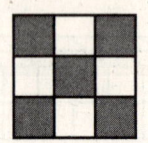

6.

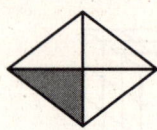

_____   _____   _____

7.

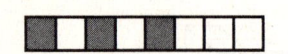

8.

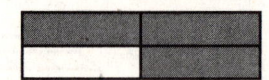

_____   _____

Draw a rectangle with the fraction shaded.

9. $\frac{1}{3}$

10. $\frac{4}{5}$

11. $\frac{5}{7}$

12. $\frac{4}{8}$

13. $\frac{4}{9}$

14. $\frac{5}{6}$

118

Use with Grade 4, Chapter 21, Lesson 1, pages 494–495.

Name_____

Parts of a Group

P 21-2 PRACTICE

Write a fraction that names what part is shaded.

1. ● ○ ○ ○

_____

2. (6 squares, 5 shaded)

_____

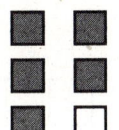

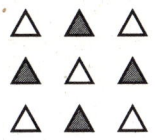

3. (9 triangles, 4 shaded)

_____

4. (8 circles, 2 shaded)

_____

5. (6 triangles, 3 shaded)

_____

6. (5 circles, 1 shaded)

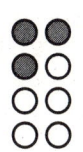

_____

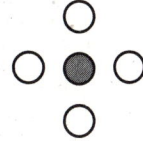

Draw a picture and then write a fraction.

7. Six of eleven balloons are blue.

_____

8. Four of seven hats have stars.

_____

9. All of five kittens are smiling.

_____

10. One of four animals is a chimpanzee.

_____

Problem Solving
Solve.

11. Five of 12 students are in the school chorus. What part of the students are in the chorus?

_____

12. Twenty of 25 students voted for class president. What part of the class did **not** vote for president?

_____

Use with Grade 4, Chapter 21, Lesson 2, pages 496–497.

Name _____

Find Equivalent Fractions and Fractions in Simplest Form

P PRACTICE 21-3

Write an equivalent fraction for each.

1. [$\frac{1}{2}$]

   _____

2. [$\frac{1}{4}$ | $\frac{1}{4}$ | $\frac{1}{4}$]

   _____

3. [$\frac{1}{5}$ | $\frac{1}{5}$ | $\frac{1}{5}$ | $\frac{1}{5}$]

   _____

Complete to find equivalent fractions.

4. $\dfrac{4 \div 2}{10 \div \square} = \dfrac{2}{\square}$

5. $\dfrac{1 \times \square}{2 \times 8} = \dfrac{\square}{16}$

6. $\dfrac{2 \div 2}{8 \div \square} = \dfrac{1}{\square}$

7. $\dfrac{1 \times \square}{5 \times 4} = \dfrac{\square}{20}$

8. $\dfrac{4}{5} = \dfrac{\square}{10}$

9. $\dfrac{1}{2} = \dfrac{6}{\square}$

10. $\dfrac{4}{\square} = \dfrac{1}{4}$

11. $\dfrac{9}{12} = \dfrac{\square}{4}$

Name an equivalent fraction for each.

12. $\dfrac{3}{7} = $ ____

13. $\dfrac{4}{5} = $ ____

14. $\dfrac{6}{15} = $ ____

15. $\dfrac{4}{12} = $ ____

Write each fraction in simplest form.

16. $\dfrac{4}{10} = $ ____

17. $\dfrac{6}{12} = $ ____

18. $\dfrac{3}{18} = $ ____

19. $\dfrac{6}{18} = $ ____

20. $\dfrac{8}{12} = $ ____

21. $\dfrac{3}{21} = $ ____

22. $\dfrac{10}{30} = $ ____

23. $\dfrac{8}{20} = $ ____

24. $\dfrac{5}{15} = $ ____

25. $\dfrac{9}{24} = $ ____

26. $\dfrac{12}{24} = $ ____

27. $\dfrac{24}{32} = $ ____

Algebra Complete the pattern of equivalent fractions.

28. $\dfrac{1}{4} = \dfrac{\square}{8} = \dfrac{\square}{12} = \dfrac{\square}{16} = \dfrac{\square}{20} = \dfrac{\square}{24}$

29. $\dfrac{1}{3} = \dfrac{\square}{6} = \dfrac{\square}{9} = \dfrac{\square}{12} = \dfrac{\square}{15} = \dfrac{\square}{18}$

Problem Solving
Solve.

30. A box contains 6 red pencils and 8 black pencils. What fraction of the pencils are red?

31. Paul caught 9 bass and 3 trout. What fraction of the fish were trout?

© Macmillan/McGraw-Hill. All rights reserved.

Use with Grade 4, Chapter 21, Lesson 3, pages 498–501.

Name_____

Compare and Order Fractions • Algebra

21-4 PRACTICE

Complete. Write >, <, or =.

1. $\frac{1}{2}$ ◯ $\frac{1}{3}$
2. $\frac{2}{5}$ ◯ $\frac{2}{7}$
3. $\frac{4}{9}$ ◯ $\frac{2}{3}$
4. $\frac{2}{5}$ ◯ $\frac{3}{4}$
5. $\frac{7}{10}$ ◯ $\frac{4}{5}$
6. $\frac{3}{4}$ ◯ $\frac{2}{3}$
7. $\frac{4}{5}$ ◯ $\frac{12}{15}$
8. $\frac{1}{5}$ ◯ $\frac{4}{20}$
9. $\frac{1}{5}$ ◯ $\frac{2}{15}$
10. $\frac{5}{12}$ ◯ $\frac{1}{4}$
11. $\frac{3}{4}$ ◯ $\frac{13}{16}$
12. $\frac{8}{9}$ ◯ $\frac{7}{8}$
13. $\frac{7}{12}$ ◯ $\frac{5}{6}$
14. $\frac{3}{10}$ ◯ $\frac{4}{9}$
15. $\frac{7}{8}$ ◯ $\frac{3}{4}$
16. $\frac{9}{10}$ ◯ $\frac{4}{5}$
17. $\frac{1}{4}$ ◯ $\frac{5}{16}$
18. $\frac{3}{5}$ ◯ $\frac{7}{10}$

Order from least to greatest.

19. $\frac{1}{4}, \frac{1}{2}, \frac{1}{5}$ ____, ____, ____
20. $\frac{7}{8}, \frac{3}{4}, \frac{3}{8}$ ____, ____, ____
21. $\frac{5}{7}, \frac{1}{7}, \frac{3}{21}$ ____, ____, ____
22. $\frac{4}{9}, \frac{1}{3}, \frac{2}{3}$ ____, ____, ____

Order from greatest to least.

23. $\frac{1}{2}, \frac{2}{3}, \frac{3}{4}$ ____, ____, ____
24. $\frac{4}{9}, \frac{2}{9}, \frac{5}{9}$ ____, ____, ____
25. $\frac{1}{4}, \frac{3}{4}, \frac{3}{16}$ ____, ____, ____
26. $\frac{5}{6}, \frac{7}{12}, \frac{3}{4}$ ____, ____, ____

Problem Solving

Solve.

27. Sandra eats $\frac{1}{6}$ of a cake. Pat eats $\frac{1}{3}$ of the same cake. Who eats more cake? Explain.

28. Karl eats $\frac{1}{2}$ of a pizza. Tim eats $\frac{2}{3}$ of a pizza. Chris eats $\frac{3}{4}$ of a pizza. Order the amounts from greatest to least.

_____

Use with Grade 4, Chapter 21, Lesson 4, pages 502–504.

Name_____

Problem Solving: Skill
Check for Reasonableness

21-5 PRACTICE

Solve.

1. There are 32 rides at an amusement park. Norman goes on $\frac{3}{8}$ of the rides. How many rides does he go on?

   _____

2. Donna took 18 rides. She went on the roller coaster $\frac{2}{3}$ of the time. How many roller-coaster rides did Donna take?

   _____

3. A dozen students go to the amusement park. A group of $\frac{1}{3}$ of these students goes on the Super Cycle. How many students go on the Super Cycle?

   _____

4. There were 25 students at the amusement park. Of these students, $\frac{2}{5}$ were there for the first time. How many students were there for the first time?

   _____

5. Each car of the Sling Shot can hold 15 people. A car is $\frac{2}{5}$ full. How many people are in the car?

   _____

6. An amusement park has 36 rides. Bobby goes on $\frac{1}{2}$ of them. How many rides does he go on?

   _____

7. Ashley puts 45 stamps in an album. She puts the same number of stamps on each page, and 3 stamps on the last page. There are 2 more pages in the album than the number of stamps on each page. How many pages are in the album? How many stamps are on each page?

   _____
   _____
   Strategy: _____

8. In the 4th grade at Spring Lake School, 189 students have pet cats, 203 students have dogs, and 83 students have cats and dogs. Make a Venn diagram to show this information.

   _____
   Strategy: _____

122

Use with Grade 4, Chapter 21, Lesson 5, pages 506–507.

Name _____

Explore Finding Parts of a Group

P 21-6
PRACTICE

Use the models to help you find the fraction of each group.

1.

$\frac{1}{3}$ of 6 = _____

2.

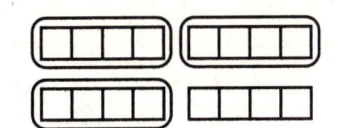

$\frac{3}{4}$ of 16 = _____

3.

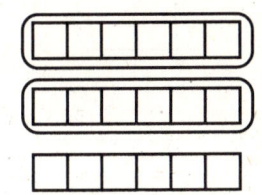

$\frac{2}{3}$ of 18 = _____

4.

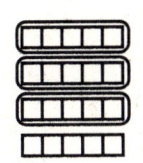

$\frac{3}{4}$ of 20 = _____

5.

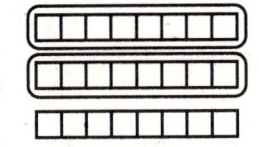

$\frac{2}{3}$ of 24 = _____

6.

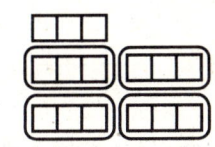

$\frac{4}{5}$ of 15 = _____

Find the fraction of each number. You may use connecting cubes.

7. $\frac{1}{2}$ of 18 = _____
8. $\frac{2}{3}$ of 15 = _____
9. $\frac{3}{5}$ of 30 = _____
10. $\frac{5}{6}$ of 12 = _____
11. $\frac{3}{7}$ of 14 = _____
12. $\frac{1}{8}$ of 32 = _____
13. $\frac{2}{9}$ of 18 = _____
14. $\frac{1}{10}$ of 40 = _____
15. $\frac{4}{7}$ of 21 = _____
16. $\frac{5}{8}$ of 40 = _____
17. $\frac{1}{3}$ of 21 = _____
18. $\frac{1}{4}$ of 20 = _____
19. $\frac{2}{5}$ of 30 = _____
20. $\frac{1}{6}$ of 36 = _____
21. $\frac{3}{8}$ of 16 = _____
22. $\frac{3}{7}$ of 28 = _____
23. $\frac{6}{7}$ of 49 = _____
24. $\frac{7}{10}$ of 60 = _____

Problem Solving
Solve.

25. Of the 24 fourth graders in Mrs. Williams' class, $\frac{1}{4}$ participate in sports. How many fourth-grade students participate in sports?

_____

26. Steven practices cello 15 hours a week. On Monday he practices $\frac{1}{5}$ of that time. How many hours does Steven practice cello on Monday?

_____

Use with Grade 4, Chapter 21, Lesson 6, pages 508–509.

Name _____

Mixed Numbers

P 21-7 PRACTICE

Rename in simplest form. If already in simplest form, rename as a mixed number.

1. $\frac{8}{7}$ = _____
2. $\frac{9}{2}$ = _____
3. $\frac{7}{2}$ = _____
4. $\frac{10}{3}$ = _____

5. $6\frac{2}{6}$ = _____
6. $3\frac{6}{8}$ = _____
7. $4\frac{1}{5}$ = _____
8. $1\frac{5}{7}$ = _____

9. $\frac{22}{10}$ = _____
10. $\frac{21}{6}$ = _____
11. $\frac{13}{2}$ = _____
12. $\frac{19}{4}$ = _____

13. $5\frac{2}{6}$ = _____
14. $2\frac{2}{8}$ = _____
15. $3\frac{2}{6}$ = _____
16. $8\frac{3}{4}$ = _____

17. $\frac{40}{6}$ = _____
18. $\frac{30}{4}$ = _____
19. $\frac{64}{6}$ = _____
20. $\frac{48}{5}$ = _____

Algebra Use the number line to compare. Write >, <, or =.

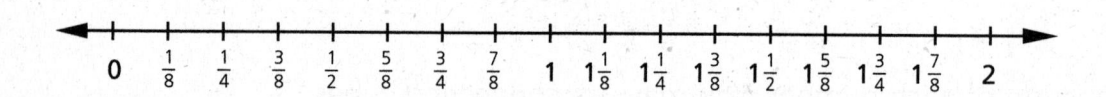

0 $\frac{1}{8}$ $\frac{1}{4}$ $\frac{3}{8}$ $\frac{1}{2}$ $\frac{5}{8}$ $\frac{3}{4}$ $\frac{7}{8}$ 1 $1\frac{1}{8}$ $1\frac{1}{4}$ $1\frac{3}{8}$ $1\frac{1}{2}$ $1\frac{5}{8}$ $1\frac{3}{4}$ $1\frac{7}{8}$ 2

21. $1\frac{1}{6}$ ◯ $1\frac{1}{8}$
22. 1 ◯ $\frac{8}{8}$
23. 2 ◯ $1\frac{7}{8}$

24. $1\frac{1}{4}$ ◯ $1\frac{5}{8}$
25. $1\frac{1}{8}$ ◯ $1\frac{1}{2}$
26. $1\frac{3}{4}$ ◯ $1\frac{7}{8}$

Problem Solving
Solve.

27. Ben measures ten one-fourths of a cup of water. What is this as a mixed number?

_____

28. Claudia ran $4\frac{1}{3}$ miles on Monday. On Tuesday she ran $4\frac{1}{2}$ miles. On which day did Claudia run a longer distance? Explain.

_____

29. Jared drank $\frac{7}{4}$ cups of juice. Aida drank $\frac{9}{6}$ cups. Who drank more juice? Explain.

_____

30. Mary worked $8\frac{1}{2}$ hours on Monday and $8\frac{3}{5}$ hours on Tuesday. On which day did she work longer? Explain.

_____

124

Use with Grade 4, Chapter 21, Lesson 7, pages 510–512.

Name _____

Problem Solving: Strategy
Draw a Tree Diagram

P 22-3 PRACTICE

Use a tree diagram to solve.

1. You spin a spinner with 4 equal sections marked 1–4. Then you spin another spinner with 3 equal sections colored red, blue, and yellow. What are all of the possible outcomes?

   _____
   _____

2. Karen throws a dart at a target with 5 equal sections marked 1–5. She then throws a dart at a target with two equal sections colored green and blue. What are all of the possible outcomes?

   _____
   _____

3. The Boardwalk Shop sells souvenir shirts. The shirts come with long sleeves or short sleeves. The shirts come in white, gray, and blue. What are all of the different kinds of shirts?

   _____
   _____

4. Boardwalk Burgers sells burgers made from beef, turkey, chicken, or soy. Burgers can have no cheese, Swiss cheese, or American cheese. How many different choices are there?

   _____
   _____

Mixed Strategy Review
Solve. Use any strategy.

5. The Target Toss Game has 6 rings. The first ring is worth 4 points, the second ring is worth 8 points, and the third ring is worth 12 points. If the pattern continues, what is the sixth ring worth?

   _____

   Strategy: _____

6. **Social Studies** In a recent year, $\frac{11}{100}$ of all U.S. vacations included time at the beach, $\frac{6}{100}$ included time at sports events, and $\frac{8}{100}$ included time at theme parks. Write these activities in order from least to most popular.

   _____

   Strategy: _____

7. Marnie brought $75 to the amusement park. She has $39 left. How much money did Marnie spend?

   _____

   Strategy: _____

8. **Create a problem** which can be solved by drawing a tree diagram. Share it with others.

   _____

Use with Grade 4, Chapter 22, Lesson 3, pages 524–525.

Name _____

Explore Making Predictions

P 22-4 PRACTICE

Use the spinner for problems 1–6.

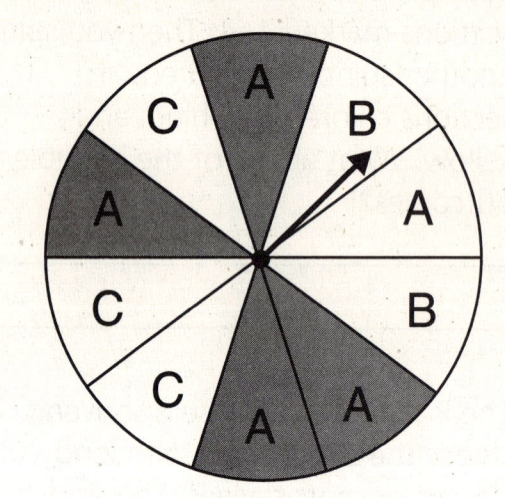

1. If you spin the spinner 100 times, what is the probability you will land on A?

   _____

2. If you spin the spinner 50 times, what is the probability you will land on B?

   _____

3. If you spin the spinner 100 times, what is the probability you will land on C?

   _____

4. If you spin the spinner 100 times, what is the probability you will land on a shaded section?

   _____

5. If you spin the spinner 50 times, what is the probability you will land on an unshaded section?

   _____

6. If you spin the spinner 50 times, what is the probability you will land on an A or a B?

   _____

Use a number cube with the faces labeled 1–6 for problems 7–10.

7. Predict the number of times 3 will come up if you toss the number cube 30 times.

   _____

8. If you toss the number cube 60 times, how often might 4 come up?

   _____

9. Is it reasonable to predict that you will toss a 4 on the number cube 2 out 12 tosses?

   _____

10. Can you predict exactly how many times 5 will come up when you toss a number cube labeled 1–6?

    _____

Use with Grade 4, Chapter 22, Lesson 4, pages 526–527.

Name _____

Add Fractions with Unlike Denominators

P PRACTICE 23-3

Add. Write each sum in simplest form.

1. $\frac{1}{4} + \frac{1}{8}$
2. $\frac{2}{3} + \frac{3}{6}$
3. $\frac{8}{12} + \frac{3}{4}$
4. $\frac{5}{6} + \frac{1}{3}$
5. $\frac{1}{5} + \frac{2}{15}$
6. $\frac{1}{6} + \frac{3}{12}$

7. $\frac{1}{6} + \frac{2}{3}$
8. $\frac{1}{3} + \frac{6}{15}$
9. $\frac{1}{2} + \frac{6}{10}$
10. $\frac{1}{2} + \frac{5}{6}$
11. $\frac{1}{2} + \frac{7}{8}$
12. $\frac{3}{5} + \frac{7}{10}$

13. $\frac{1}{2} + \frac{1}{4} =$ ____
14. $\frac{3}{10} + \frac{1}{2} =$ ____
15. $\frac{1}{6} + \frac{5}{12} =$ ____

16. $\frac{1}{4} + \frac{3}{8} =$ ____
17. $\frac{3}{12} + \frac{2}{3} =$ ____
18. $\frac{1}{5} + \frac{5}{15} =$ ____

19. $\frac{3}{4} + \frac{3}{8} =$ ____
20. $\frac{7}{9} + \frac{1}{3} =$ ____
21. $\frac{1}{4} + \frac{5}{12} =$ ____

22. $\frac{10}{12} + \frac{3}{4} =$ ____
23. $\frac{1}{2} + \frac{5}{6} + \frac{1}{3} =$ ____
24. $\frac{1}{8} + \frac{1}{2} + \frac{3}{4} =$ ____

Algebra Compare. Write >, <, or =.

25. $\frac{1}{4} + \frac{9}{12}$ ◯ $\frac{1}{4} + \frac{2}{3}$
26. $\frac{2}{6} + \frac{1}{6}$ ◯ $\frac{1}{2} + \frac{1}{4}$

27. $\frac{2}{12} + \frac{1}{4}$ ◯ $\frac{3}{12} + \frac{1}{6}$
28. $\frac{3}{5} + \frac{4}{10}$ ◯ $\frac{1}{2} + \frac{2}{10}$

Problem Solving
Solve.

29. Your family ate $\frac{1}{2}$ of a box of cereal one day and $\frac{3}{4}$ the next. Did your family eat more or less than 1 box of cereal? Explain.

_____

30. At 6:00 P.M., $\frac{1}{6}$ of the passengers boarded the plane. At 6:10 P.M., $\frac{2}{3}$ of the passengers boarded. What fraction of the passengers are on the plane?

_____

Use with Grade 4, Chapter 23, Lesson 3, pages 546–548.

Name_____

Problem Solving: Skill
Choose an Operation

P 23-4 PRACTICE

Solve.

1. Max buys $\frac{7}{8}$ pound of apples and $\frac{3}{8}$ pound of grapes. What is the total amount of fruit he buys?

   _____

2. Adela makes 20 cookies. She gives 15 cookies to her friends. What part of the 20 cookies is left?

   _____

3. Chen buys $\frac{5}{8}$ pound of American cheese and $\frac{1}{4}$ pound of Swiss cheese. How much more American cheese than Swiss cheese does he buy?

   _____

4. Kathryn uses $\frac{3}{4}$ tablespoon of nutmeg and $\frac{3}{4}$ tablespoon of cocoa. How many tablespoons of nutmeg and cocoa does she use altogether?

   _____

5. Amy buys $\frac{1}{4}$ pound of turkey and $\frac{1}{4}$ pound of honey-roasted ham. How much meat does she buy altogether?

   _____

6. Marge cuts a cherry pie into 8 slices. She eats one slice. What part of the pie is left?

   _____

7. A recipe for pudding uses $\frac{3}{10}$ cup of milk. A recipe for custard uses $\frac{2}{5}$ cup of milk. How much milk do both recipes use?

   _____

8. Patrick bought $\frac{3}{4}$ pound of large cookies and $\frac{1}{12}$ pound of small cookies. What is the total weight of the cookies he bought?

   _____

Mixed Strategy Review
Solve. Use any strategy.

9. Jamal has 30 coins. He has 5 more nickels than dimes and 5 fewer quarters than dimes. How many of each kind of coin does Jamal have?

   _____

   Strategy: _____

10. Caroline needs to arrive at school at 8:45 A.M. It takes her 10 minutes to get dressed, 10 minutes to eat breakfast, and 15 minutes to walk to school. At what time should Caroline get up?

    _____

    Strategy: _____

Name_____

Subtract Fractions with Like Denominators   **P** 24-1 PRACTICE

Subtract. Write each difference in simplest form.

1. $\frac{4}{5} - \frac{2}{5}$
2. $\frac{5}{7} - \frac{3}{7}$
3. $\frac{5}{8} - \frac{1}{8}$
4. $\frac{8}{9} - \frac{2}{9}$
5. $\frac{5}{6} - \frac{1}{6}$
6. $\frac{4}{9} - \frac{1}{9}$

7. $\frac{7}{10} - \frac{2}{10}$
8. $\frac{6}{10} - \frac{4}{10}$
9. $\frac{7}{12} - \frac{1}{12}$
10. $\frac{4}{15} - \frac{1}{15}$
11. $\frac{8}{11} - \frac{4}{11}$
12. $\frac{11}{12} - \frac{8}{12}$

13. $\frac{7}{9} - \frac{2}{9} =$ ____
14. $\frac{5}{16} - \frac{1}{16} =$ ____
15. $\frac{7}{8} - \frac{3}{8} =$ ____

16. $\frac{5}{7} - \frac{4}{7} =$ ____
17. $\frac{8}{9} - \frac{1}{9} =$ ____
18. $\frac{4}{5} - \frac{3}{5} =$ ____

19. $\frac{7}{12} - \frac{5}{12} =$ ____
20. $\frac{7}{12} - \frac{4}{12} =$ ____
21. $\frac{10}{11} - \frac{5}{11} =$ ____

22. $\frac{11}{12} - \frac{8}{12} =$ ____
23. $\frac{9}{10} - \frac{5}{10} =$ ____
24. $\frac{7}{8} - \frac{3}{8} =$ ____

25. $\frac{2}{3} - \frac{2}{3} =$ ____
26. $\frac{8}{9} - \frac{2}{9} =$ ____
27. $\frac{9}{11} - \frac{8}{11} =$ ____

Algebra Compare. Write >, <, or =.

28. $\frac{5}{9} - \frac{2}{9} \bigcirc \frac{6}{9} - \frac{3}{9}$
29. $\frac{7}{10} - \frac{3}{10} \bigcirc \frac{8}{10} - \frac{2}{10}$

30. $\frac{5}{12} - \frac{1}{12} \bigcirc \frac{7}{12} - \frac{5}{12}$
31. $\frac{11}{15} - \frac{10}{15} \bigcirc \frac{14}{15} - \frac{13}{15}$

32. $\frac{7}{11} - \frac{6}{11} \bigcirc \frac{7}{11} - \frac{5}{11}$
33. $\frac{12}{13} - \frac{5}{13} \bigcirc \frac{9}{13} - \frac{2}{13}$

Problem Solving
Solve.

34. At lunch you cut a sandwich into 4 parts and eat 3 of the parts. What fraction of the sandwich is left?

_____

35. For breakfast and lunch you drink $\frac{2}{3}$ of a quart of milk. How much of the quart is left?

_____

Use with Grade 4, Chapter 24, Lesson 1, pages 556–557.

Name_____

Explore Subtracting Fractions with Unlike Denominators

P 24-2 PRACTICE

Subtract. Write each difference in simplest form.

1. $\frac{1}{4} - \frac{1}{8}$

    $\frac{2}{8} - \frac{1}{8} =$ ___

2. $\frac{2}{3} - \frac{1}{6}$

    $\frac{4}{6} - \frac{1}{6} =$ ___

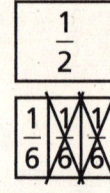

3. $\frac{1}{2} - \frac{1}{3}$

    $\frac{3}{6} - \frac{2}{6} =$ ___

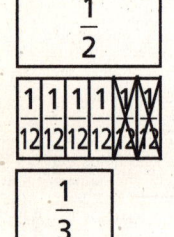

4. $\frac{1}{2} - \frac{2}{12}$

    $\frac{6}{12} - \frac{2}{12} =$ ___

5. $\frac{1}{5} - \frac{1}{10}$

    $\frac{2}{10} - \frac{1}{10} =$ ___

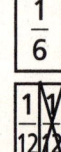

6. $\frac{1}{6} - \frac{1}{12}$

    $\frac{2}{12} - \frac{1}{12} =$ ___

7. $\frac{3}{12} - \frac{1}{6} =$ ___

8. $\frac{1}{2} - \frac{2}{10} =$ ___

9. $\frac{1}{4} - \frac{1}{12} =$ ___

10. $\frac{7}{12} - \frac{1}{3} =$ ___

11. $\frac{7}{9} - \frac{2}{3} =$ ___

12. $\frac{5}{12} - \frac{1}{4} =$ ___

13. $\frac{5}{6} - \frac{1}{3} =$ ___

14. $\frac{3}{4} - \frac{4}{12} =$ ___

15. $\frac{1}{2} - \frac{1}{12} =$ ___

16. $\frac{1}{2} - \frac{3}{10} =$ ___

17. $\frac{5}{6} - \frac{1}{12} =$ ___

18. $\frac{1}{2} - \frac{3}{8} =$ ___

134   Use with Grade 4, Chapter 24, Lesson 2, pages 558–559.

Name _____

Subtract Fractions with Unlike Denominators   **P** 24-3 PRACTICE

Subtract. Write each difference in simplest form.

1. $\frac{1}{3} - \frac{1}{12}$
2. $\frac{3}{4} - \frac{5}{12}$
3. $\frac{1}{5} - \frac{2}{15}$
4. $\frac{7}{10} - \frac{1}{5}$
5. $\frac{11}{12} - \frac{5}{6}$
6. $\frac{5}{6} - \frac{2}{3}$

7. $\frac{9}{10} - \frac{3}{5}$
8. $\frac{3}{4} - \frac{1}{2}$
9. $\frac{3}{5} - \frac{3}{10}$
10. $\frac{5}{9} - \frac{1}{3}$
11. $\frac{2}{3} - \frac{2}{9}$
12. $\frac{3}{4} - \frac{1}{8}$

13. $\frac{5}{8} - \frac{1}{4} =$ ___
14. $\frac{2}{3} - \frac{1}{6} =$ ___
15. $\frac{1}{4} - \frac{1}{12} =$ ___

16. $\frac{4}{5} - \frac{7}{10} =$ ___
17. $\frac{4}{9} - \frac{1}{3} =$ ___
18. $\frac{4}{5} - \frac{3}{10} =$ ___

19. $\frac{1}{2} - \frac{1}{6} =$ ___
20. $\frac{3}{8} - \frac{1}{4} =$ ___
21. $\frac{7}{9} - \frac{1}{3} =$ ___

22. $\frac{7}{12} - \frac{1}{6} =$ ___
23. $\frac{1}{2} - \frac{1}{4} =$ ___
24. $\frac{2}{3} - \frac{5}{12} =$ ___

25. $\frac{7}{12} - \frac{1}{2} =$ ___
26. $\frac{7}{10} - \frac{2}{5} =$ ___
27. $\frac{1}{2} - \frac{2}{10} =$ ___

Algebra Find each missing number.

28. $\frac{7}{8} - \frac{1}{2} = \frac{\square}{8}$
29. $\frac{5}{6} - \frac{1}{\square} = \frac{2}{3}$
30. $\frac{3}{4} - \frac{\square}{12} = \frac{2}{3}$

31. $\frac{3}{6} - \frac{1}{\square} = \frac{1}{3}$
32. $\frac{2}{3} - \frac{1}{6} = \frac{\square}{2}$
33. $\frac{5}{9} - \frac{1}{\square} = \frac{2}{9}$

Problem Solving
Solve.

34. Pam has $\frac{7}{8}$ yard of ribbon. She uses $\frac{1}{2}$ yard. How much ribbon does Pam have left?

35. Joe has $\frac{5}{6}$ yard of fabric. He uses $\frac{2}{3}$ yard to make a kite. How much fabric does Joe have left?

Name _____

Problem Solving: Strategy
Solve a Simpler Problem

P 24-4 PRACTICE

Solve using a simpler problem.

1. Sandwiches cost $4.95. Drinks cost $0.99. How much does it cost to buy 2 sandwiches and 3 drinks?

   _____

2. A customer pays $3.95 for 5 pounds of apples. What is the price for 1 pound of apples?

   _____

3. Recipe A uses $\frac{1}{2}$ cup of chicken broth and $\frac{1}{4}$ cup of water. Recipe B uses $\frac{1}{3}$ cup of chicken broth and $\frac{1}{3}$ cup of water. Which recipe uses more liquid?

   _____

4. Tracy buys $\frac{3}{4}$ pound of roast beef, $\frac{1}{2}$ pound of turkey, and $\frac{3}{8}$ pound of ham. Ken buys $\frac{1}{4}$ pound of roast beef, $\frac{1}{2}$ pound of turkey, and $\frac{3}{8}$ pound of ham. Who buys more meat? How much more does that person buy?

   _____

Mixed Strategy Review
Solve. Use any strategy.

5. There are 24 plants in a garden. There are 4 more tomato plants than red pepper plants. There are twice as many red pepper plants as green pepper plants. How many of each kind of plant is in the garden?

   _____
   _____

   Strategy: _____

6. The Yogurt Cart has the following 3 flavors: chocolate, vanilla, and strawberry. Yogurt comes in a cup or a cone. You can have no sprinkles, chocolate sprinkles, or rainbow sprinkles. How many different choices are there?

   _____
   _____

   Strategy: _____

7. **Health** An ounce of cheddar cheese has 114 calories. An ounce of Brie cheese has 95 calories. How many more calories does an ounce of cheddar cheese have than an ounce of Brie cheese?

   _____

   Strategy: _____

8. **Write a problem** that you will use a simpler problem to solve. Share it with others.

   _____
   _____
   _____

136

Use with Grade 4, Chapter 24, Lesson 4, pages 564–565.

Name _____

Circle Graphs

P 24-5 PRACTICE

What does one section in each circle graph shown below represent?

1.   2.   3.

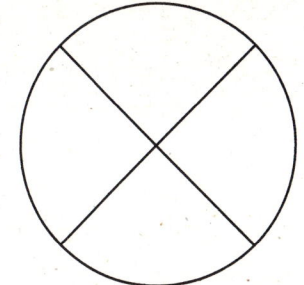

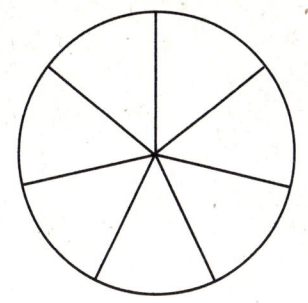

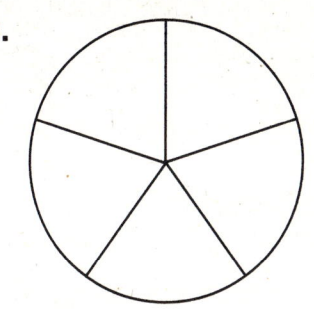

_____   _____   _____

Use the circle graph. Solve the problems.

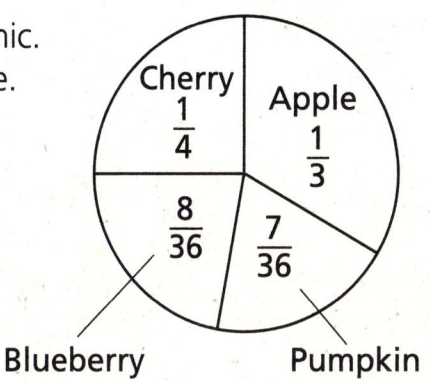

Lana surveyed 36 people who were going to a picnic. She asked them to choose their favorite type of pie. She made a circle graph of the data.

Favorite Pie

Cherry $\frac{1}{4}$, Apple $\frac{1}{3}$, $\frac{8}{36}$ Blueberry, $\frac{7}{36}$ Pumpkin

4. How many people chose cherry?

   _____

5. List the choices from most to least popular.

   _____

6. Which type of pie did 12 people choose?

   _____

Sam sold 24 sandwiches at the school fair. He made a circle graph to show the sales.

7. Complete the circle graph to find the number of chicken sandwiches sold.

   _____

8. How many egg salad sandwiches were sold?

   _____

Favorite Sandwich

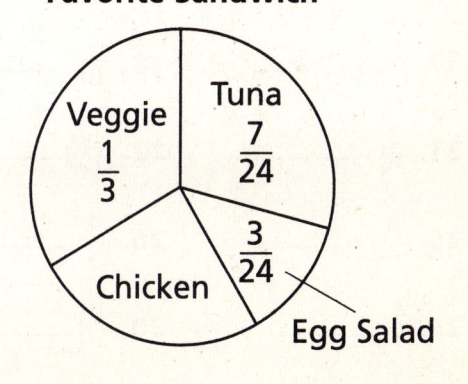

Veggie $\frac{1}{3}$, Tuna $\frac{7}{24}$, Chicken, Egg Salad $\frac{3}{24}$

Use with Grade 4, Chapter 24, Lesson 5, pages 566–568.

137

Name _____

Explore Fractions and Decimals

P 25-1 PRACTICE

Write a fraction and a decimal for each shaded part. Then write the fraction in simplest form.

1.

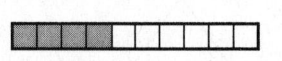

2.

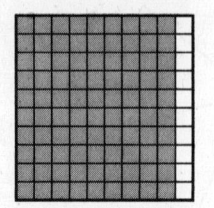

3.

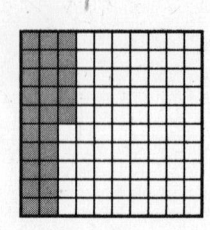

4.

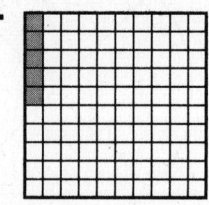

_____ _____ _____ _____

5.

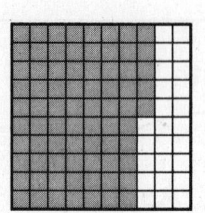

6.

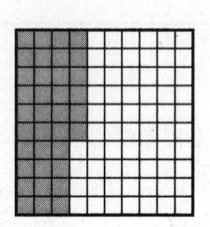

7.

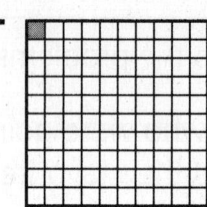

8.

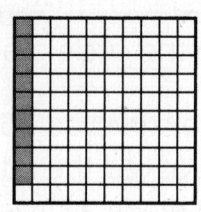

_____ _____ _____ _____

9.

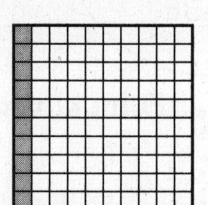

10.

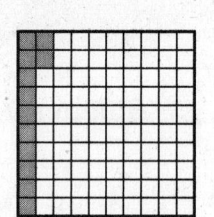

11.

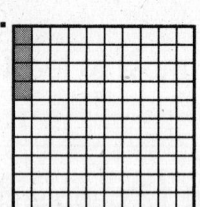

12.

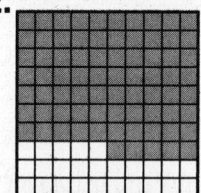

_____ _____ _____ _____

Write each fraction as a decimal.

13. $\frac{70}{100}$ _____   14. $\frac{78}{100}$ _____   15. $\frac{13}{100}$ _____   16. $\frac{27}{100}$ _____

17. $\frac{8}{10}$ _____   18. $\frac{5}{10}$ _____   19. $\frac{1}{100}$ _____   20. $\frac{4}{100}$ _____

21. $\frac{3}{10}$ _____   22. $\frac{66}{100}$ _____   23. $\frac{7}{10}$ _____   24. $\frac{90}{100}$ _____

25. $\frac{4}{10}$ _____   26. $\frac{1}{2}$ _____   27. $\frac{10}{25}$ _____   28. $\frac{5}{20}$ _____

29. $\frac{4}{5}$ _____   30. $\frac{10}{50}$ _____   31. $\frac{3}{4}$ _____   32. $\frac{2}{5}$ _____

138

Use with Grade 4, Chapter 25, Lesson 1, pages 584–585.

Name_____

Tenths and Hundredths

P 25-2 PRACTICE

Write a fraction and a decimal for each part that is shaded. Then write the fraction in simple form.

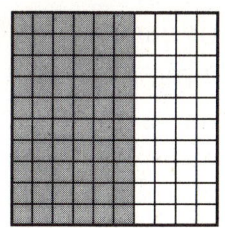

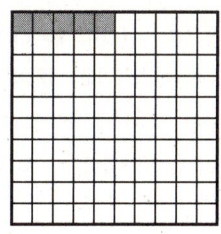

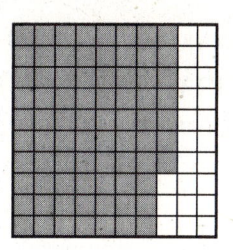

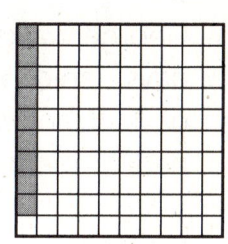

1. _____   2. _____   3. _____   4. _____

Write each as a decimal.

5. $\frac{2}{5}$ _____   6. $\frac{7}{10}$ _____   7. $\frac{1}{4}$ _____   8. $\frac{7}{100}$ _____

9. $\frac{1}{2}$ _____   10. $\frac{1}{10}$ _____   11. $\frac{2}{100}$ _____   12. $\frac{96}{100}$ _____

13. two tenths _____   14. fifteen hundredths _____

15. six hundredths _____   16. three tenths _____

17. five tenths _____   18. seventeen hundredths _____

19. ninety-nine hundredths _____   20. two tenths _____

Write a fraction and a decimal for each point. Tell if it is closer to 0, $\frac{1}{2}$, or 1.

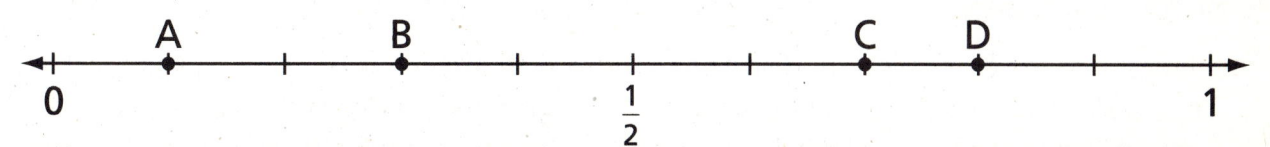

21. A _____   22. B _____

23. C _____   24. D _____

Problem Solving
Solve.

25. Peter's house is 0.78 mile from school. Write the number in words.

26. Lora walks for five tenths of an hour. Write the number as a decimal.

_____   _____

Use with Grade 4, Chapter 25, Lesson 2, pages 586–588.

Name _____

Problem Solving: Skill
Choose a Representation

P 25-3
PRACTICE

Choose a representation and solve.

1. George walks to work 6 out of 10 days. Janice walks to work 0.7 of 10 days. Who walks to work a greater part of the time?

   _____

2. Train Q is on time or early 0.4 of the time. Train Y is on time or early $\frac{1}{2}$ of the time. Which train is on time or early a lesser part of time?

   _____

3. In a survey, 0.5 of the people who answer say that they are very satisfied with subway service. Four tenths of the people say that they are somewhat satisfied. Are more people very satisfied or somewhat satisfied?

   _____

4. Colleen takes the bus 18 of the days in June. Rita takes the bus $\frac{7}{10}$ of the days in June. Who takes the bus more days? [HINT: June has 30 days.]

   _____

5. Alfredo walks to work 12 out of 20 days. He says he walks to work 0.9 of those days. Is his statement reasonable?

   _____

6. The express bus is late 0.2 of the time. A reporter says that the express bus is late $\frac{2}{10}$ of the time. Is the reporter's statement reasonable?

   _____

Mixed Strategy Review

Solve. Use any strategy.

7. Tamara collected shells. On Monday she collected 3 shells. On Tuesday she collected 6 shells, on Wednesday 12 shells, and on Thursday 24 shells. How many shells did Tamara collect on Sunday?

   _____

   Strategy: _____

8. Richard has a red shirt, a yellow shirt, and a white shirt. He has black pants, brown pants, and blue pants. What are all the possible outfits Richard can make?

   _____
   _____
   _____
   _____
   _____

   Strategy: _____

Use with Grade 4, Chapter 25, Lesson 3, pages 590–591.

Name_____

Thousandths

P 25-4 PRACTICE

Write each as a decimal.

1. $\frac{123}{1,000}$ _____
2. $\frac{370}{1,000}$ _____
3. $\frac{25}{1,000}$ _____
4. $\frac{4}{1,000}$ _____

5. $\frac{17}{1,000}$ _____
6. $\frac{225}{1,000}$ _____
7. $\frac{36}{1,000}$ _____
8. $\frac{1}{1,000}$ _____

9. $\frac{6}{1,000}$ _____
10. $\frac{24}{1,000}$ _____
11. $\frac{3}{1,000}$ _____
12. $\frac{12}{1,000}$ _____

13. $\frac{120}{1,000}$ _____
14. $\frac{999}{1,000}$ _____
15. $\frac{9}{1,000}$ _____
16. $\frac{60}{1,000}$ _____

17. sixteen thousandths _____
18. twenty-five thousandths _____
19. nine thousandths _____
20. three hundred twenty-nine thousandths _____
21. five hundred thousandths _____
22. six hundred ninety thousandths _____
23. ninety-five thousandths _____
24. two thousandths _____
25. eleven thousandths _____
26. four thousandths _____
27. seventy-two thousandths _____
28. one hundred ninety-nine thousandths _____

Algebra Complete.

29.

meters	decimeters	centimeters	millimeters
_____	0.06	0.6	6
0.009	_____	_____	_____
_____	_____	_____	14

Problem Solving
Solve.

30. Joe weighs 0.625 g of rice. Write this amount in words.

_____

_____

31. Jaime bats three hundred one thousandths for the season. Write this as a decimal.

_____

Use with Grade 4, Chapter 25, Lesson 4, pages 592–593.

Name_____

Decimals Greater Than 1

P 26-1 PRACTICE

Write as a mixed number in simplest form and a decimal to tell how much is shaded.

1. 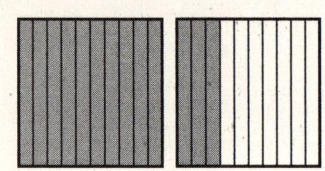   2. 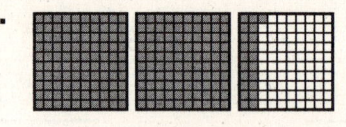   3.

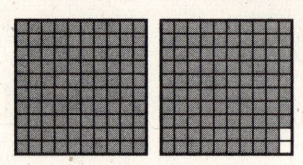

_____   _____   _____

Write the fraction as a decimal.

4. $7\frac{3}{10}$ 5. $1\frac{25}{100}$ 6. $9\frac{5}{100}$ 7. $8\frac{125}{1,000}$

_____   _____   _____   _____

8. $6\frac{2}{100}$ 9. $17\frac{7}{10}$ 10. $8\frac{5}{1,000}$ 11. $3\frac{37}{1,000}$

_____   _____   _____   _____

12. $9\frac{1}{10}$ 13. $2\frac{9}{10}$ 14. $27\frac{21}{100}$ 15. $25\frac{16}{1,000}$

_____   _____   _____   _____

16. $18\frac{98}{100}$ 17. $13\frac{5}{1,000}$ 18. $10\frac{12}{1,000}$ 19. $11\frac{3}{100}$

_____   _____   _____   _____

20. $6\frac{6}{100}$ 21. $19\frac{375}{1,000}$ 22. $23\frac{8}{10}$ 23. $76\frac{60}{1,000}$

_____   _____   _____   _____

24. $24\frac{4}{100}$ 25. $11\frac{1}{100}$ 26. $9\frac{19}{100}$ 27. $6\frac{26}{100}$

_____   _____   _____   _____

28. eight and three tenths   29. seven and seventy hundredths

_____   _____

Problem Solving

Solve.

30. Out of 100 pairs of shoes in a sporting goods store, 53 pairs are running shoes. What decimal shows the number of pairs of running shoes?

_____

31. Out of 1,000 backpacks, 25 are red and the rest are green. What decimal shows the number of red backpacks?

_____

Use with Grade 4, Chapter 26, Lesson 1, pages 598–599.

Compare and Order Decimals • Algebra

26-2 PRACTICE

Compare. Write >, <, or =.

1. 0.2 ◯ 0.02   2. 0.7 ◯ 0.70   3. 1.78 ◯ 1.87   4. 12.16 ◯ 12.160
5. 0.106 ◯ 0.160   6. 5.117 ◯ 5.107   7. 11.99 ◯ 12.1   8. 11.1 ◯ 10.1
9. 9.06 ◯ 9.16   10. 6.5 ◯ 5.9   11. 2.1 ◯ 0.2   12. 10.3 ◯ 10.300
13. 16.75 ◯ 16.57   14. 14.44 ◯ 14.54   15. 18.01 ◯ 18.11   16. 9.1 ◯ 9.09
17. 21.12 ◯ 22.13   18. 16.06 ◯ 16.6   19. 1.1 ◯ 1.11   20. 9.9 ◯ 10.0
21. 9.01 ◯ 9.10   22. 14.03 ◯ 13.99   23. 2.22 ◯ 2.11   24. 19.99 ◯ 18.99

Write in order from greatest to least.

25. 1.78, 1.08, 1.87   26. 0.88, 0.08, 0.98   27. 1.11, 1.21, 0.22

28. 10.02, 9.9, 10.12   29. 7.7, 8.8, 7.07   30. 1.001, 1.011, 1.111

Write in order from least to greatest.

31. 0.01, 0.1, 0.001   32. 2.22, 2.02, 2.12   33. 6.07, 5.99, 6.17

34. 1.06, 1.16, 0.99   35. 11.17, 10.99, 9.99   36. 16.6, 16.61, 16.1

Problem Solving
Solve.

37. On Monday Ken ran 100 meters in 11.2 seconds. On Tuesday he ran 100 meters in 10.9 seconds. On which day did Ken run faster?

38. Jadwin Bridge is 1.6 km long. Seely Bridge is 1.06 km long. Which bridge is longer?

Use with Grade 4, Chapter 26, Lesson 2, pages 600–602.

Name _____

Problem Solving: Strategy
Draw a Diagram

P 26-3 PRACTICE

Draw a diagram to solve.

1. CD World is 1.8 miles east of the school. William lives 1.4 miles west of the school. Sound City is 2.9 miles east of William. Is William closer to CD World or to Sound City?

   _____

2. Silver Hills is 3.9 miles north of Bay Edge. East Ridge is 1.3 miles south of Silver Hills. East Ridge is 2.8 miles north of Hightown. How far is Bay Edge from Hightown?

   _____

3. Ed walks up 2 floors from his office to the storeroom. He walks down 6 floors to the cafeteria. How many floors away is the cafeteria from Ed's office?

   _____

4. A cab driver leaves his garage. He goes north 9 blocks, south 6 blocks, and north 8 blocks. How many blocks is he from his garage?

   _____

Mixed Strategy Review
Solve. Use any strategy.

5. The City Sports Center offers season tickets, 20-game tickets, or single game tickets. Seats are available for the lower deck, middle deck, or upper deck. You can buy an individual seat or a pair of seats. How many choices do you have?

   _____

   Strategy: _____

6. **Social Studies** In 1996, Abilene, Texas, had a population of 122,130. Amarillo, Texas, had a population that was 83,885 greater than the population of Abilene. What was the population of Amarillo?

   _____

   Strategy: _____

7. There are 48 people at a dinner at City Hall. You want to use small tables that seat 5 people and large tables that seat 8 people. To have full tables, which tables should be used? How many of these tables will be needed?

   _____

   Strategy: _____

8. **Write a problem** which you could solve by drawing a diagram. Share it with others.

   _____
   _____
   _____

144

Use with Grade 4, Chapter 26, Lesson 3, pages 604–605.

Name_____

Round Decimals

26-4 PRACTICE

Round to the nearest whole number.

1. 9.47 ____
2. 2.8 ____
3. 6.01 ____
4. 9.09 ____

5. 1.1 ____
6. 3.51 ____
7. 4.62 ____
8. 1.5 ____

9. 13.61 ____
10. 25.09 ____
11. 37.8 ____
12. 52.4 ____

13. 93.56 ____
14. 88.48 ____
15. 19.71 ____
16. 63.44 ____

Round to the nearest tenth.

17. 7.24 ____
18. 9.43 ____
19. 6.58 ____
20. 8.89 ____

21. 3.25 ____
22. 1.27 ____
23. 3.98 ____
24. 7.24 ____

25. 31.26 ____
26. 71.64 ____
27. 12.55 ____
28. 64.93 ____

29. 47.96 ____
30. 87.54 ____
31. 29.69 ____
32. 36.97 ____

33. 53.84 ____
34. 19.46 ____
35. 61.07 ____
36. 78.85 ____

Round to the nearest hundredth.

37. 8.236 ____
38. 4.186 ____
39. 9.275 ____
40. 1.123 ____

41. 6.008 ____
42. 2.055 ____
43. 7.266 ____
44. 3.199 ____

45. 17.246 ____
46. 26.981 ____
47. 78.006 ____
48. 91.115 ____

49. 53.102 ____
50. 66.666 ____
51. 32.333 ____
52. 45.999 ____

53. 13.462 ____
54. 51.277 ____
55. 90.409 ____
56. 45.555 ____

Problem Solving

Solve.

57. A vitamin pill weighs 2.346 g. What is its mass to the nearest hundredth of a gram?

_____

58. Jason weighs 152.6 lb. What is his weight to the nearest pound?

_____

Use with Grade 4, Chapter 26, Lesson 4, pages 606–608.

Name _____

Explore Adding Decimals

P 27-1
PRACTICE

Use the models to find each sum.

1.

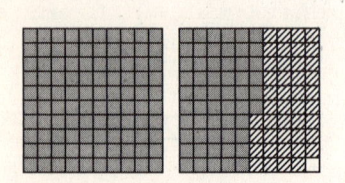

1.56 + 0.43 = _____

2.

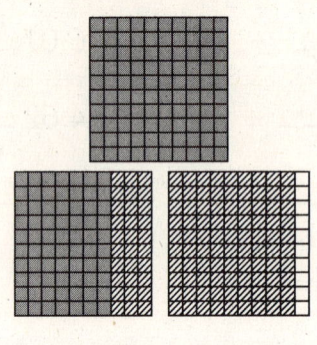

1.7 + 1.2 = _____

3.

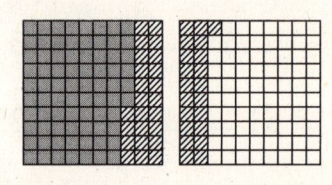

0.76 + 0.45 = _____

Find each sum.

4.	0.3 + 0.4	**5.**	0.5 + 0.4	**6.**	0.6 + 0.7	**7.**	0.8 + 0.9	**8.**	0.4 + 0.6
9.	0.99 + 0.88	**10.**	0.62 + 0.53	**11.**	0.71 + 0.59	**12.**	0.44 + 0.79	**13.**	0.86 + 0.13
14.	2.7 + 3.8	**15.**	0.5 + 1.9	**16.**	2.6 + 1.8	**17.**	1.7 + 2.8	**18.**	0.4 + 0.9

19. 0.85 + 2.17 = _____ **20.** 2.76 + 1.32 = _____ **21.** 3.46 + 1.78 = _____

22. 2.96 + 2.23 = _____ **23.** 0.67 + 2.98 = _____ **24.** 0.12 + 2.2 = _____

25. 1.5 + 2.49 = _____ **26.** 2.14 + 1.9 = _____ **27.** 2.3 + 1.92 = _____

Problem Solving
Solve.

28. Two strips of paper, 3.6 cm long and 2.8 cm long, are taped together. How long is the entire strip of paper?

29. One apple has a mass of 0.26 kg. Another apple has a mass of 0.87 kg. What is the mass of both apples?

Name_____

Add Decimals

27-2 Practice

Add.

1. 0.36 + 0.25	2. 0.29 + 0.44	3. 0.60 + 0.70	4. 1.67 + 1.45	5. 2.67 + 1.38
6. 4.2 + 6.4	7. 1.2 + 8.3	8. 0.697 + 9.262	9. 23.604 + 5.408	10. 32.75 + 12.30
11. 25.97 + 0.12	12. 12.32 + 1.74	13. 13.407 + 26.708	14. 21.151 + 4.774	15. 6.373 + 5.602
16. 2.874 + 8.129	17. 36.215 + 9.759	18. 12.948 + 7.267	19. 0.254 + 12.259	20. 3.187 + 6.975
21. 11.3 6.7 + 21.6	22. 8.25 4.30 + 9.20	23. 4.142 8.167 + 2.94	24. 4.567 13.621 + 21.984	25. 7.0 9.288 +12.6

26. 12.5 + 11.35 = _____   27. 2.7 + 2.73 = _____   28. 3.36 + 5.031 = _____

29. 3.869 + 9.3 + 7.76 = _____   30. 7.35 + 8.2 + 17.314 = _____

31. 12.42 + 7.687 + 19.3 = _____   32. 8.0 + 4.343 + 10.5 = _____

Algebra Find the number you need to add to complete the pattern. Then complete.

33. 1.3, 1.9, 2.5, _____, _____, _____   Add _____

34. 4.12, 4.125, _____, 4.135, _____, _____   Add _____

Problem Solving
Solve.

35. Lora spends $2.64 on stamps and $1.39 on envelopes. How much does she spend?

36. Ben buys packing tape for $2.97 and boxes for $6.99. How much does he spend?

Use with Grade 4, Chapter 27, Lesson 2, pages 626–628.

Problem Solving: Skill
Choose an Operation

Solve.

1. The train trip from Springfield to Morris Hill is 6.2 miles. The next stop, Peapack, is 3.2 miles from Morris Hills. How long is the train trip from Springfield to Peapack?

   _____

2. The train trip from Point Dume to Snug Harbor is 8.31 miles. The road from Point Dume to Snug Harbor is 9.6 miles. How much longer is the road trip than the train?

   _____

3. Daniel biked 6.24 miles last week. This week he biked 1.65 miles less than last week. How far did he bike this week?

   _____

4. Myra bikes 3.25 miles from home to the record store. Then she bikes 1.1 miles to the movie theater. How many miles does she bike altogether?

   _____

5. Eddie rode 1.9 miles more today than he did yesterday. He rode 5.75 miles yesterday. How far did Eddie ride today?

   _____

6. Shore Road is 6.3 miles long. Nicole has biked 2.2 miles along Shore Road so far. How many miles does she have left?

   _____

Mixed Strategy Review
Solve. Use any strategy.

7. Maritza has a penny and a number cube with faces numbered in multiples of 10 from 10–60. What are all the possible outcomes of tossing the number cube and flipping the penny?

   _____

   _____

   Strategy: _____

8. Bill, Jack, and Marla are fourth graders. Bill is older than Marla and Jack. Jack is younger than Marla. What is the order of the fourth graders from oldest to youngest?

   _____

   Strategy: _____

Name _____

Estimate Sums

P 27-4 PRACTICE

Estimate. Round to the nearest whole number.

1. 5.1 + 9.4 _____
2. 6.7 + 8.4 _____
3. 1.9 + 3.8 _____
4. $6.35 + $5.95 _____
5. 7.45 + 8.56 _____
6. 4.32 + 7.59 _____
7. 9.3 + 2.6 _____
8. 22.63 + 3.46 _____
9. 31.06 + 9.98 _____
10. 45.92 + 4.18 _____
11. $33.19 + $9.50 _____
12. 6.67 + 21.15 _____

Add. Estimate to check for reasonableness.

13. 19.76 + 9.55 = _____
14. $10.25 + $3.25 = _____
15. 19.67 + 9.94 = _____
16. 3.7 + 5.2 + 4.6 = _____
17. 4.1 + 9.6 + 1.9 = _____
18. 2.9 + 6.7 + 7.3 = _____
19. $3.75 + $9.90 + $8.75 = _____
20. 4.76 + 9.15 + 8.95 = _____
21. 8.12 + 4.79 + 7.15 = _____
22. $6.30 + $7.95 + $8.10 = _____
23. 7.75 + 8.90 + 9.90 = _____
24. 2.178 + 6.472 + 8.015 = _____

Algebra Compare. Write > or <.

25. 3.7 + 2.5 ◯ 1.9 + 4.2
26. 4.9 + 1.6 ◯ 5.1 + 3.1
27. 6.9 + 7.1 ◯ 3.8 + 8.3
28. 9.2 + 3.6 ◯ 2.6 + 9.1
29. 5.5 + 6.3 ◯ 8.2 + 5.2
30. 9.4 + 2.7 ◯ 6.8 + 6.1
31. 1.6 + 2.9 ◯ 3.1 + 1.1
32. 7.7 + 7.2 ◯ 8.1 + 9.1
33. 8.7 + 9.6 ◯ 9.1 + 8.6

Problem Solving
Solve.

34. The odometer on a new car shows 17.7 miles. Sean drives the car 12.9 miles. About how many miles does the odometer show now?

35. Lenny buys one CD for $12.75 and another CD for $18.90. About how much does Lenny pay for the two CDs?

_____   _____

Use with Grade 4, Chapter 27, Lesson 4, pages 632–633.

Name _____

Choose a Computation Method

P 27-5 PRACTICE

Add. Tell which method you used.

1. 20.9
  + 7.4
  ———

  Method: _____

2. $47.53
  + 26.87
  ———

  Method: _____

3. 377.91
  + 45.66
  ———

  Method: _____

4. 7,135.9
  + 1,255.3
  ———

  Method: _____

5. 245.066
  + 327.814
  ———

  Method: _____

6. $295.88
  + 407.67
  ———

  Method: _____

7. 37.8 + 1,245.9 = _____

  Method: _____

8. $639.75 + $59.05 = _____

  Method: _____

9. 4,275.84 + 1,362.45 = _____

  Method: _____

Algebra Choose two addends from the box to complete each number sentence.

10. _____ + _____ = $43.77

11. _____ + _____ = $45.89

$24.66	$25.03
$20.86	$19.11

Problem Solving
Solve.

12. Arnie has a large hamburger, a soft drink, and a piece of pie for lunch every day. The hamburger costs $3.75, the drink costs $0.85, and the piece of pie costs $1.95. How much does Arnie pay for 5 lunches?

  _____

13. The cook where Arnie eats lunch uses 110.5 grams of ground beef in each small hamburger and 124.8 grams in each large hamburger. What is the total mass of beef used in 2 small and 1 large hamburger?

  _____

Name _____

Explore Subtracting Decimals

P 28-1 PRACTICE

Use the models to find each difference.

1. 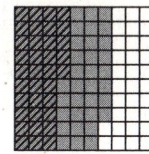 0.68 − 0.35 = _____
2. 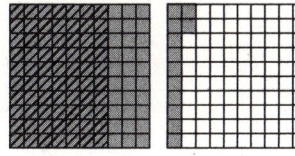 1.12 − 0.7 = _____
3. 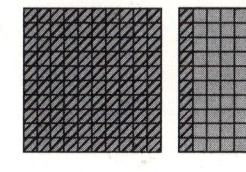 1.8 − 1.1 = _____

Find each difference.

4. 0.9 − 0.3
5. 1.2 − 0.6
6. 2.7 − 0.9
7. 2.5 − 1.6
8. 2.1 − 1.7

9. 1.67 − 0.48
10. 1.6 − 1.48
11. 3.11 − 1.12
12. 3.7 − 2.91
13. 1.2 − 1.13

14. 3.6 − 1.47
15. 2.02 − 1.79
16. 0.95 − 0.67
17. 0.8 − 0.25
18. 0.74 − 0.59

19. 1.7 − 0.35
20. 2.04 − 1.69
21. 1.03 − 0.6
22. 0.80 − 0.54
23. 2.0 − 1.06

24. 2.7 − 1.6 = _____
25. 0.8 − 0.5 = _____
26. 7.66 − 2.34 = _____
27. 1.52 − 0.57 = _____
28. 0.73 − 0.57 = _____
29. 0.70 − 0.34 = _____
30. 0.8 − 0.07 = _____
31. 0.4 − 0.14 = _____
32. 3.7 − 0.16 = _____

Problem Solving
Solve.

33. A board is 2.12 m long. A piece 1.55 m long is cut from it. How much of the board is left?

34. A piece of wire is 2.6 cm long. A piece 1.9 cm long is cut from it. How much of the wire is left?

Use with Grade 4, Chapter 28, Lesson 1, pages 640–641.

Name _____

Subtract Decimals

P 28-2 PRACTICE

Subtract. Check each answer.

1. 0.7 − 0.4	2. 6.3 − 0.7	3. 9.1 − 2.3	4. 4.5 − 2.7	5. 1.2 − 0.7	6. 0.43 − 0.26
7. 0.44 − 0.22	8. 7.04 − 3.66	9. 15.03 − 3.12	10. 4.12 − 1.27	11. 9.00 − 0.09	12. 7.17 − 2.70
13. 9.04 − 7.50	14. 6.00 − 4.70	15. 8.20 − 4.96	16. 5.34 − 4.67	17. 1.67 − 0.50	18. 19.83 − 3.60
19. 8.154 − 2.075	20. 17.076 − 0.027	21. 5.258 − 3.129	22. 8.000 − 2.974	23. 1.755 − 0.896	24. 6.024 − 2.402

25. 6.7 − 2.4 = _____
26. 7.6 − 2.07 = _____
27. 8.5 − 3.08 = _____
28. 9.03 − 3.775 = _____
29. 7.44 − 3.867 = _____
30. 4.627 − 2.88 = _____
31. 3.6 − 2.79 = _____
32. 8.36 − 3.248 = _____
33. 4.556 − 0.93 = _____
34. 34.0 − 2.097 = _____

Algebra Find each missing number.

35. $7.97 - n = 0.52$ _____
36. $h - 4.64 = 2.31$ _____
37. $5.25 + b = 10.46$ _____
38. $a + 7.08 = 18.5$ _____

Problem Solving
Solve.

39. Christine buys a pair of socks for $8.35. What is her change from a $10 bill?

_____

40. Matt buys a pencil for $0.35, a pen for $2.75, and a ruler for $4.36. What is his change from a $20 bill?

_____

Use with Grade 4, Chapter 28, Lesson 2, pages 642–644.

Name _____

Problem Solving: Strategy
Act It Out

P 28-3 PRACTICE

Act it out to solve.

1. The tennis team travels to a statewide contest. They buy 8 student bus tickets at $6.95 each and 2 adult bus tickets at $9.50 each. How much does the team spend for tickets?

   _____

2. A bus ticket costs $8.75. A train ticket for the same ride costs $12.50. Suppose you buy 4 tickets. How much money would you save by taking the bus instead of the train?

   _____

3. A bus driver earns $16.40 per hour for the first 7 hours of work each day. She earns $24.60 per hour for each hour over 7 hours. How much does she earn in a 9-hour day?

   _____

4. The Silver Eagle Express has a dining car. Sandwiches cost $5.95. Drinks cost $1.49. How much does a family pay for 3 sandwiches and 4 drinks?

   _____

Mixed Strategy Review
Solve. Use any strategy.

5. Sam spends $18.40 on a train ticket, $5.90 on a cab, and $11.20 on dinner. He has $30 left. How much money did Sam have when he started?

   _____
   Strategy: _____

6. **Science** The first steam-powered railroad engine was built in England in 1804. Thomas Edison tested an electric-powered railroad engine 76 years later. When did Edison test his engine?

   _____
   Strategy: _____

7. Teri has 17 model trains. She has a long shelf that can hold 7 trains. She also has 2 smaller shelves. How can she arrange the trains on shelves so that each smaller shelf has an equal number of trains?

   _____
   Strategy: _____

8. **Write a problem** which you could act out to help you find the answer. Share it with others.

   _____
   _____
   _____

Use with Grade 4, Chapter 28, Lesson 3, pages 646–647.

Name _____

Estimate Differences

P 28-4 PRACTICE

Estimate. Round to the nearest whole number.

1. 6.3 − 2.6 _____
2. 7.1 − 4.8 _____
3. 8.7 − 5.2 _____
4. 9.0 − 3.9 _____
5. 4.6 − 1.5 _____
6. 7.34 − 5.78 _____
7. 8.57 − 3.52 _____
8. 17.26 − 13.78 _____
9. 26.14 − 12.95 _____
10. $34.95 − $12.20 _____
11. 25.60 − 11.55 _____
12. 47.15 − 17.11 _____

Subtract. Estimate to check for reasonableness.

13. 7.1 − 2.70 = _____
14. 9.8 − 4.6 = _____
15. 8.5 − 6.3 = _____
16. 5.6 − 1.75 = _____
17. 36.62 − 23.13 = _____
18. 24.35 − 10.4 = _____
19. 77.36 − 15.93 = _____
20. $16.12 − $12.80 = _____
21. 94.32 − 22.80 = _____
22. $54.10 − $34.89 = _____
23. 13.4 − 6.79 = _____
24. 47.65 − 17.93 = _____
25. $14.75 − $6.90 = _____
26. 63.5 − 18.27 = _____

Algebra Compare. Write > or <.

27. 7.2 − 3.5 ◯ 8.8 − 5.4
28. 9.9 − 4.8 ◯ 6.4 − 1.7
29. 7.6 − 2.2 ◯ 5.6 − 1.3
30. 8.3 − 6.6 ◯ 4.2 − 2.3
31. 9.1 − 8.7 ◯ 2.1 − 1.1
32. 7.2 − 4.5 ◯ 6.8 − 5.8
33. 5.2 − 2.3 ◯ 9.7 − 7.9
34. 9.3 − 3.8 ◯ 9.9 − 3.1
35. 8.1 − 4.6 ◯ 7.2 − 5.1

Problem Solving

Solve.

36. Jake has $25.75. He spends $13.15 on magazines. About how much money does Jake have left?

_____

37. Nancy ran a total of 5.7 miles today. She ran 3.2 miles this morning. About how many miles did Nancy run this afternoon?

_____

154 Use with Grade 4, Chapter 28, Lesson 4, pages 648–649.

Name _____

Choose a Computation Method

P 28-5 PRACTICE

Solve. Tell which method you used.

1.   21.9
  − 7.45
  ———

  Method: _____

2. $48.13
  − 26.87
  ———

  Method: _____

3.   307.61
  − 45.96
  ———

  Method: _____

4.   35.92
  − 15.31
  ———

  Method: _____

5.   0.81
  − 0.52
  ———

  Method: _____

6. $95.38
  − 47.67
  ———

  Method: _____

7. $37.8 - 25\frac{3}{10} =$ _____
  Method: _____

8. $71.4 - 33.7 =$ _____
  Method: _____

9. $\$63.75 - \$59.55 =$ _____
  Method: _____

10. $5\frac{3}{10} - 2.1 =$ _____
  Method: _____

11. $16\frac{7}{10} - 10.5 =$ _____
  Method: _____

12. $\$82.30 - 47.14 =$ _____
  Method: _____

Algebra Choose a number from the box to complete each number sentence.

13. 53.75 − _____ = $25.14

14. 50.81 − _____ = $22.67

| $28.14 |
| $28.61 |

Problem Solving

Solve.

15. Sue lives 27.21 kilometers from Ramon. She lives 36.8 kilometers from Zack. How much further from Zack than Ramon does Sue live?

_____

16. Bob's computer is 32.1 centimeters long. Juanita's computer is 31.6 centimeters long. How much longer is Bob's computer than Juanita's computer?

_____

Use with Grade 4, Chapter 28, Lesson 5, pages 650–652.

Summer Skills Refresher

Summer Skills

Playing in the Parks

Many of Florida's state parks have upland areas for picnicking and playing. Submerged areas are used for swimming, boating, snorkeling, and scuba diving.

1. Jon Pennekamp Coral Reef State Park is one of the most popular parks for viewing coral and underwater wildlife. There are 60,124.31 submerged acres in the park. Write the number in expanded form.

   _____

2. Pennekamp State Park also includes 2,960.22 upland acres. Write the number of acres in word form.

   _____

3. How many more submerged acres than upland acres are there at Pennekamp State Park?

   _____

4. Honeymoon Island State Park offers 1,507.13 acres of submerged park. How many more submerged acres are in Pennekamp State Park than in Honeymoon Island State Park?

   _____

Answers: 1. 60,000 + 100 + 20 + 4 + 0.3 + 0.01; 2. Two thousand, nine hundred sixty and twenty-two hundredths; 3. 57,164.09 acres; 4. 58,617.18 acres

Summer Skills 159

5. Sebastian Inlet State Park has 726.54 acres of upland park and 143.14 acres of submerged park. How many more acres of upland park are there?

_____

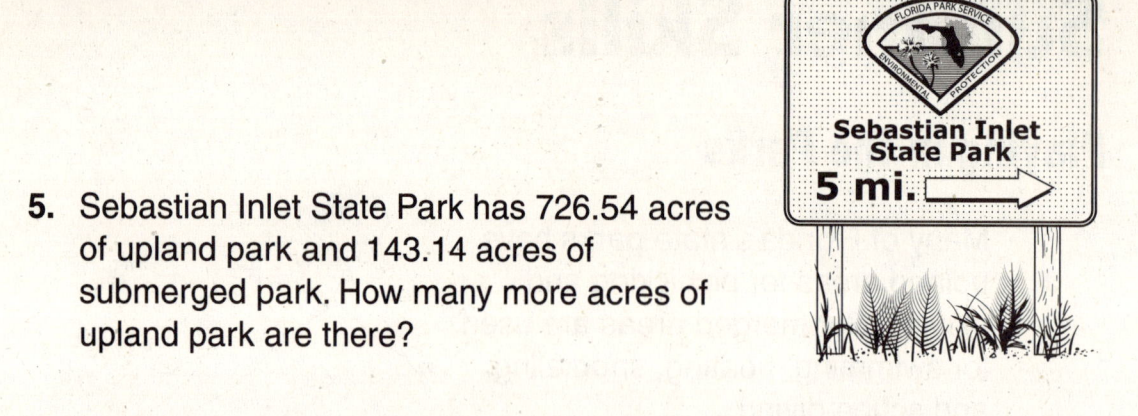

6. Write the number of upland acres at Sebastian Inlet in word form. Then write it in expanded form.

_____

_____

7. How many upland acres are in Sebastian Inlet and Pennekamp state parks combined?

_____

8. How many submerged acres are in Sebastian Inlet and Pennekamp state parks combined?

_____

9. About how many times more submerged acres are there than upland acres in the two parks combined?

_____

Answers: 5. 583.4 acres; 6. Seven hundred twenty-six and fifty-four hundredths 700 + 20 + 6 + 0.5 + 0.04; 7. 3,686.76 acres; 8. 60,267.45 acres; 9. about 15 times more

Summer Skills

Summer Skills

Recipe

Jodi made a picnic lunch to take to Ocean World. She made Florida crab cakes using the ingredients below:

Florida Crab Cakes

1 pound crabmeat

2 shallots minced

$\frac{1}{2}$ medium red bell pepper

$\frac{1}{2}$ medium green bell pepper

3 tablespoons minced parsley

$\frac{1}{4}$ cup dry bread crumbs

$1\frac{1}{2}$ teaspoons seafood seasoning

$\frac{1}{4}$ teaspoon ground pepper

1. The crabmeat was packaged in ounces. How many ounces of crabmeat should Jodi buy?

2. Jodi lost her $\frac{1}{4}$-cup measuring cup. She knows that one cup is equal to 16 tablespoons. How many tablespoons of breadcrumbs should Jodi use?

3. Three teaspoons equals 1 tablespoon. How many teaspoons of minced parsley should Jodi use?

4. Jodi also wants to make a salad for her family. Should she measure the salad oil in cups or gallons?

Answers: 1. 16 ounces; 2. 4 tablespoons; 3. 9 teaspoons; 4. cups

Ocean World Animals

These animals can be found at Ocean World.

Animal	Average Length
American Alligator	3.3 meters
Sea Turtle	121 cm
Aldabra Tortoise	122 cm
Carpet Python	2.4 m
Nile Crocodile	5 m

5. What is the length, in centimeters, of the Nile crocodile?

6. What is the length, in meters, of the sea turtle?

7. How many more centimeters is the Nile crocodile than the Aldabra tortoise?

8. How many more centimeters is the American alligator than the carpet python?

Answers: 5. 500 cm; 6. 1.21 m; 7. 378 cm; 8. 90 cm

Summer Skills

Happy Land Park

Florida has many amusement parks. Happy Land Park offers rides and shows for children and adults. Look carefully at the park's sign.

1. Name two closed shapes used in the Happy Land sign.

   _____

2. Suppose the radius of the Ferris wheel at the park is 50 feet. What is the diameter of the Ferris wheel?

   _____

3. Which two rides have a circle in their shapes?

   _____

4. Suppose the park's entrance sign is 10 feet wide by 15 feet long. What is the area of the sign?

   _____

Answers: 1. possible answer: circle, square; 2. 100 feet; 3. Ferris wheel, carousel; 4. 150 square feet

Summer Skills

5. Sketch a reflection of the park entrance in the box below.

6. When the sign is reflected, which of the shapes are congruent?

_____

_____

7. Suppose you created a reflection of the Ferris wheel. Would both pictures look exactly the same? Explain your thinking.

_____

_____

8. If the entrance sign were reflected, what letters would appear exactly the same?

_____

_____

Answers: **5.** check drawings; **6.** all shapes will be congruent; **7.** no, the wheel is flattened and not fully round like a circle; **8.** H, A, Y

Summer Skills

Planting Trees

The Broward County Brownie Troop celebrated Arbor Day by planting trees for the chamber of commerce. Each troop planted 5 trees. There are 60 Brownie troops in Broward County.

1. Write an expression to represent the number of trees planted.

2. Use your expression from problem 1 to find the number of trees planted.

3. Suppose 10 more Brownie troops also planted trees. Change the expression from problem 1 to show this new information.

4. Use the expression from problem 3 to find the new number of trees planted.

Answers: 1. 5 × 60; 2. 300 trees; 3. 5 × 70; 4. 350 trees

Day Camp

At Gator Day Camp Challenge, campers are grouped in teams. Campers are given colored shirts to wear during the games; each team wears a different color.
Use the table to answer problems 5–8.

Number of Teams	2	4	6	8
Number of Shirts Needed	14	28	42	63

5. What is the relationship between the number of teams and the number of shirts needed?

   _____

6. Write an expression to represent the number of shirts needed for 6 teams.

   _____

7. How many shirts would be needed for 5 teams?

   _____

8. Suppose the number of campers per team changed to 8. Fill in the table below using this information.

Number of Teams	2	4	6	8
Number of Shirts Needed				

Answers: 5. multiply by 7; 6. 6 × 7; 7. 35 shirts; 8. 16; 32; 48; 64

Summer Skills

Head, Heart, Hands, and Health

The Timpoochee 4-H Center in Niceville, Florida offers summer educational programs for children of all ages. One of the most popular programs is the Challenge Adventure Course. Working through the maze of ropes, platforms, trees, and logs helps children develop self-confidence and increases their coordination and physical ability.

1. Thomas was spending a week at the 4-H center. During that time, he climbed the Challenge Adventure Course every day. His times are shown in the table below. Finish the line graph to show Thomas's progress in this section of the course.

Day 1	53 sec.
Day 2	58 sec.
Day 3	46 sec.
Day 4	49 sec.
Day 5	52 sec.
Day 6	47 sec.
Day 7	45 sec.

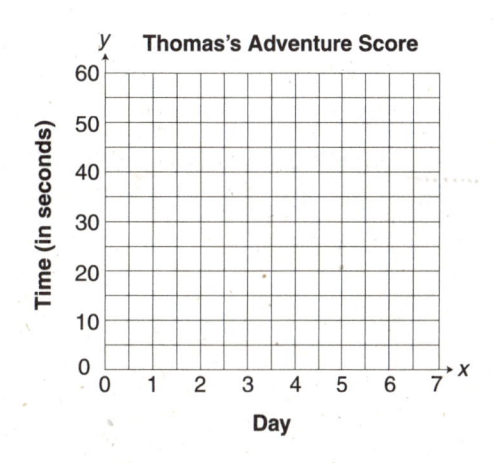

2. How did Thomas's score change during the week?

   _____

3. What is the range of Thomas's scores?

   _____

Answers: 1. Check graph.; 2. Possible answer: His score went up and down a little during the week, but overall it dropped.; 3. 13 seconds

Center Scheduling

The 4-H center schedules activity periods throughout the day. Campers move from activity to activity so they are able to practice many skills.

4. The clocks below show the starting time and the ending time for the first activity period. How long is the first activity period?

   _____

5. The clock below shows the time the second activity period starts. Campers use the time in between periods to move from one activity area to another. How long do the campers have to move between the first and second activity periods?

   _____

6. Each night, the camp plays *Taps* on the PA system. Every morning, the camp plays *Reveille*. The clocks below show the times for *Taps* and *Reveille*. How much time passes between the two bugle calls?

   Taps   Reveille

   _____

Answers: 4. 1 hour, 15 minutes; 5. 10 minutes; 6. 9 hours, 30 minutes